CORPORATE FLEXIBILITY

Corporate Flexibility

A Comparative Analysis of Small Firms in Northern Ireland and Massachusetts

NOLA HEWITT-DUNDAS
Northern Ireland Economic Research Centre, Belfast

Avebury
Aldershot · Brookfield USA · Hong Kong · Singapore · Sydney

Published by
Avebury
Ashgate Publishing Ltd
Gower House
Croft Road
Aldershot
Hants GU11 3HR
England

Ashgate Publishing Company
Old Post Road
Brookfield
Vermont 05036
USA

British Library Cataloguing in Publication Data

Hewitt-Dundas, Nola
Corporate flexibility : a comparative analysis of small firms in Northern Ireland and Massachusetts
1.Organizational change - Northern Ireland 2.Organizational change - Massachusetts 3.Small business - Northern Ireland 4.Small business - Massachusetts
I.Title
338.6'42'09416

Library of Congress Catalog Card Number: 96-80379

ISBN 1 85972 370 5

Printed in Great Britain by the Ipswich Book Company, Suffolk

Contents

List of tables

List of figures

Preface

This book examines the rhetoric which advocates flexibility as the key to economic rejuvenation. Traditional theories of industrial location are increasingly being challenged by non-hierarchical models concerning regions' innovation systems. In the new models the traditional pre-determination of industrialization is replaced by self-determination. In particular, the small firm sector has been ascribed a pivotal role in stimulating regional development and facilitating economic rejuvenation.

The investigation presented in this book is comparative, between a low productivity region, Northern Ireland, and a high productivity region, Massachusetts. Massachusetts has moved away from a position of threatened peripheralization to that of core status, as manifest through the vibrant high-technology and tradable services sectors, entrepreneurialism and growth. In contrast, Northern Ireland has become entrapped within a spiral of deprivation with governmental assistance failing to revitalize the economic fabric.

The book examines the extent to which the principles of flexibility are evident within small manufacturing firms. Two questions are therefore addressed, namely, to what extent are disparities found in the adoption of flexibility within the small manufacturing firms in Northern Ireland and Massachusetts? And secondly, is there a positive correlation between the adoption of flexibility and company growth?

The context of the investigation is presented in Chapters One and Two. Chapter One outlines how the geography of industrialization is no longer predetermined but instead intricately linked to inherited characteristics within a region, including the behaviour of firms, the innovative milieu and regional policy. It is proposed that flexibility is vital to regional development, not only within the confines of the firm, but also within the social and economic fabric. Chapter Two examines the characteristics of the Northern Ireland and

Massachusetts economies. While the development trajectory of Northern Ireland is epitomized by a vicious circle of development, economic rejuvenation within Massachusetts is characterized by a virtuous circle of development.

Chapter Three outlines the matched plant methodology which was used in the investigation. One hundred small manufacturing firms were interviewed, with these being evenly split between Northern Ireland and Massachusetts. Seventy per cent of the sample was engineering firms with the remaining 30 per cent being food processing firms. The firms were *matched* by the main product being manufactured, employment and the age of the firm. These criteria were tightly controlled to ensure comparability between the samples.

The main findings of the investigation are presented in Chapters Four, Five, Six and Seven. Chapter Four, in examining the extent of turnover and employment growth within the sample of small firms, demonstrates that there were significant growth disparities between the small firms in Northern Ireland and those in Massachusetts. Having identified this growth disparity, Chapters Five and Six examine the extent of flexibility within the labour force of the small firms. While Chapter Five measures labour flexibility in terms of functional flexibility, Chapter Six, focuses upon numerical flexibility. In Chapter Seven, attention is directed away from labour flexibility and towards flexibility in the production process. Again significant regional disparities are found between the small manufacturing firms in the use of flexible production methods. The inter-linkages between labour flexibility, flexibility in the production process and small firm growth are synthesized in the concluding chapter, Chapter Eight. The extent to which flexibility within the small firms approached that proposed by the Flexible Specialization and Regulation schools of thought is also evaluated.

The findings within this book are the outcome of research undertaken in the Department of Geography at the Queen's University of Belfast. Funding for the research was provided by the Department of Education for Northern Ireland. The writing of this book was undertaken at the Northern Ireland Economic Research Centre, as part of the Centre's research program on Enterprise and Company Growth, supported by the Department of Economic Development.

The production of this book would not have been possible without the faithful and patient support of a number of individuals. Dr Steve Royle and Professor David Hitchens at the Queen's University of Belfast, who provided guidance and advice concerning the direction of the research. Mr Simon Bridge who provided an invaluable insight to the Northern Ireland development agencies. Professor Jonathan Haughton, Professor Michael Blim and Professor Alan Cromer at the Northeastern University provided invaluable advice and encouragement in undertaking the Massachusetts element of the research. Mr Levine of the Department of Employment and Training, Massachusetts, who comprehensively charted the historical development of the State of Massachusetts. Elaine Gordon, Ellen Moore and Anne Levi supplied precious time and expertise in formatting and checking the text. Finally Dr Stephen Roper of the Northern Ireland Economic Research Centre, whose patience, guidance and words of support have alone ensured the writing of this book. To all of these individuals I am extremely grateful. The judgments, analyses and views expressed are those of the author and are not necessarily those of the Northern Ireland Economic Research Centre nor its Committee of Management.

1 Flexibility: The key to economic regeneration?

> The Growth of Capitalism, and the renewal of its prodigious powers of expansion, is intimately connected to industry localization, territorial clustering and the periodic opening up of new industrial centres *(Walker 1989:89).*

Introduction

Economic activity is epitomized by an 'inconstant geography of capitalist industrialisation' (Storper and Walker, 1989:34). As spatial patterns are established the areas where economic activity is concentrated are known as the core, while the less prosperous areas are labeled the periphery. Within core regions, economic dynamism is found with concentrations of labour enriching the skills base and acting as a ready market for goods and services. In addition economies of both scale and agglomeration stimulate investments and lead to improvements in the regional infrastructure. Each of these factors fuse to generate comparative advantages in some areas relative to others. Conversely, the marginalising tendencies of unemployment, low wages, poorly developed infrastructure and low levels of investment combine to reinforce the economic stagnation and entrenchment of peripheral regions relative to the core.

Traditionally neo-classical thought (Robinson, 1971; Kaldor, 1972) has proposed that over time an equilibrium will be established whereby industrialisation will dictate the position of regions within the core-periphery framework. For example, economic prosperity in the core regions is reinforced as the remaining capital and labour within the periphery are attracted to the core. Marginalising tendencies within the periphery are reinforced through 'cumulative causation' as disparities between the core and periphery are accentuated. Core areas will therefore develop a *virtuous circle* of development as competitiveness breeds competitiveness and the resource base is enriched. The peripheral areas however will be entrapped within a *vicious circle* of development as their resources are depleted and their marginalising tendencies accentuated.

The product cycle theory and its associated New International Division of Labour (NIDL) and urban hierarchy model, have until recently been widely used as a means of understanding the geography of industrialisation. Changing conditions of production, technology and labour practices have challenged the traditional neo-classical notion of equilibrium and instead the notion of the flexible production complex and models of inter-regional technological relationships have emerged (Todling, 1992). Within the new theories, the geography of industrialisation is no longer predetermined, with peripheral regions being capable of overcoming and reversing their marginalising tendencies.

Central to the changing theoretical perception of geographical location is the tendency towards crisis within capitalism. It is proposed that there has been a crisis within the capitalist system during the past 30 years, and this has ushered in a new era of Neo-Fordism which has transformed the traditional patterns of industrialisation. Storper and Walker (1989) argue that Neo-classical equilibrium theory can no longer explain the dynamism of economic development. 'Massive deindustrialization of older economies such as Britain and the Northeastern United States should not have happened in an equilibrium world' (Storper and Walker, 1989:34). While some areas such as Massachusetts and the third Italy have overcome their deindustrialisation since the post war years to be reestablished as core areas, other areas such as Northern Ireland have experienced increasing peripherality.

This chapter focuses upon the changing forces which mould the geography of industrialisation and prosperity. The main characteristics of the new theories of industrial location are outlined briefly. The inherent tendency of capitalism towards crisis is illustrated along with the implications of this in the late 20th Century in terms of the prevailing forces of production, as these are central to the new theories of industrial location. Within Chapter Two, Northern Ireland is presented as a peripheral region which, faced with industrialisation in the post war years has become increasingly marginalized relative to core regions. In contrast Massachusetts, faced with similar structural changes in the post war years as Northern Ireland, has experienced an *economic miracle*. The rationale for contrasting Northern Ireland and Massachusetts is therefore presented.

Industrial location

The product cycle theory of industrial development (Kuznets, 1930; Burns, 1934) argues that there are three stages to the life-cycle of a product with each having a distinct geography. These three stages consist of development, maturity and standardisation. During the development stage of a product's life-cycle R&D and production activities will be located where there is a plentiful supply of skilled labour, markets, suppliers and support services. The urban hierarchy theory argues that these factors will only be found within areas where agglomeration economies can be obtained, such as in old industrial cores[1]. Urban centres will therefore be characterized by a concentration of product development and the production of products in the early stages of their life-cycle.

As sales begin to reach their maximum and products and production technology become standardized the level of skilled labour needed in production is reduced. Further, agglomeration economies become less important and firms may decide to transfer production to peripheral areas where costs of production are lower (Kleinknecht and Poot, 1992; Hymer, 1972; Cohen, 1981; Scott, 1983).

Product cycle theory is therefore spatially deterministic with the availability of resources dictating the location of industry. Old industrial cores or industrial agglomerations, with rich local resource environments will dominate the innovation and development stages of new products and industries. In contrast, hinterlands, or peripheral regions with poorer local resources will be the location of branch-plants or subcontracting as cheap factor resources are exploited. As firms increase their ability to be geographically mobile, the areas over which the stages of production occur have widened considerably. Walker (1989:86) suggests that this has led to 'the growing separation of specialized bits of production across wider national and international arenas' as evident through the new international division of labour.

The challenge to this deterministic model of industrial location has emanated from the findings of the flexible specialization school and interpretations of the Japanese structure of production. Criticisms of product life cycle and urban hierarchy theories have been based upon their over-simplistic view of technological change; the perceived stability of the urban hierarchy; and the changing nature of production since the 1970s which arguably renders the product life cycle and urban hierarchy theories obsolete. Each of these factors are dealt with briefly.

Firstly, technological development does not follow the same trajectory across industrial sectors and decentralization of production may only be applicable to some industries, such as those using Fordist methods of production and dependent on cost advantages for a competitive advantage (Walker, 1989). The urban hierarchy theory has also been criticized for its misunderstanding of the innovation process. The theory perceives innovation as a linear process initiated by scientific and technological change. Instead it is argued that innovation is an evolutionary process (Nelson and Winter, 1977; 1982), involving the development and application of existing knowledge (Dosi 1988), as embedded within the regions' innovative milieu, firms' characteristics and behaviour and local policy initiatives (Adaylot, 1986; Camagni, 1991; Maillat,

1991). The spatial hierarchy of industrial location is therefore undermined if innovation is an evolutionary social process dependent on networks.

Secondly, the pattern of territorial advantages may be unstable and transitory rather than having the very long-term persistence suggested by the agglomeration advantages associated with urban areas by the urban hierarchy theory. Evidence of this type of changing spatial pattern of advantages is offered by the urban-rural shift of activity which has taken place in the United Kingdom over the last three decades (Keeble, 1988; Keeble, Owens and Thompson, 1983; Townsend, 1983; Keeble and Tyler, 1995). Although typically attributed to factor cost advantages or property constraints (Fothergill and Gudgin, 1982) rather than any technological explanation, this dynamic process of industrial relocation does suggest that areas do experience changing patterns of advantage in the face of changing industrial requirements.

Finally, the product cycle and urban hierarchy theories as explanations for industrial location are becoming redundant as new technological, social and institutional forces emerge and alternative explanations for industrial location are proposed. The flexible specialization school (Sabel, 1982; Brusco and Sabel, 1983; Piore and Sabel, 1984) argues that the emergence of technologically advanced machinery is reversing the traditional deskilling thesis, which is central to the notion of standardized production being located in the periphery. Further, the agglomeration of firms within flexible industrial complexes is central to flexible specialization. The geography of flexible specialization is no longer characterized by international, national and regional hierarchies of command, but instead, by a return to self-sustaining regional economies (Amin, 1994).

While differences exist between different schools of thought concerning the nature of the emerging form of production, each point to an increasingly flexible and diversified production system and shorter product life-cycles. Particularly important in terms of urban hierarchy theory is the shortening or disappearance of the standardisation period (Oakey, 1984). This reduces the demand for standardized mass produced goods and increases the dependence on skilled shop-floor workers, R&D resources and agglomeration economies. This increased integration of the production process throughout each stage of the product life cycle means that 'neither the factory nor automated machine systems can alone cope with the manner of coordinating work across complex divisions of labour, and in this sense there is no such thing as a self-contained branch plant' (Walker, 1989:88).

Oakey (1984) draws out three implications from this changing organisation of production. Firstly, in this new mode of development (Lipietz, 1988) the situation of peripheral regions will be made more difficult as low factor costs become less important compared to agglomeration advantages. Secondly, the emerging high technology industrial sectors are likely to locate where agglomeration economies exist and there is a rich resource environment in terms of labour, markets, infrastructure and support services. Thirdly as fragmentation strategies are pursued by large firms greater opportunities will emerge for small firms to develop both interdependently of the large firms and independently within niche markets.

Although Oakey (1984) proposes that agglomeration economies will become more important this does not mean that core regions will continue to prosper at the expense of the periphery. More recent non-hierarchical models concerning regions' innovation systems challenge the traditional hierarchy of industrial location and instead propose that each region has a spatial pattern of advantages linked to the characteristics and behaviour of firms, the innovative milieu and regional policy. It is therefore the combination of these factors which will determine the technological trajectory of peripheries as independent from national and international hierarchies (Aydalot and Keeble, 1988). Within this perspective the notion of predetermined industrial location is challenged. The peripheral status of regions is no longer dictated by activity in the core, but instead by their own inherent characteristics. The dynamism of regional development is therefore a function of the resource environment. For this reason, if the peripheral status of regions such as Northern Ireland is to be reversed then it is essential that the relative competitive position is ascertained, so that comparative strengths within the environment can be exploited and weaknesses dealt with through the effective targeting of regional policy.

Theories of structural change in the economy

It is now widely accepted that major qualitative changes have been occurring in the developed world since the recession of the 1970s. From the mid 1970s to the 1980s the member states of the European Union lost three million jobs while the United States experienced a net increase of 18 million jobs. The employment losses within the European Union were attributed to Eurosclerosis as characterized by rigidities within the labour force, capital markets and government regulations (Nielsen, 1991). A range of 'binary histories' (Sayer and Walker, 1992) have been proposed to help explain the emerging capitalist relations. Flexibility has become synonymous with newness, while rigidity has become associated with the old restrictive Fordism which must be shunned if economic growth is to occur. These binary histories must be treated with caution as they merely present extremes by which past and present situations can be benchmarked. The theories which have emerged to explain contemporary structural change are also best regarded as conceptual frameworks through which regional activity can be gauged. The Flexible Specialization school, the neo-Schumpeterian school and the Regulation school are the main bodies of thought regarding the emerging form of capitalism. While there are significant differences between the theories, it is important to note that different perspectives within the schools of thought have increased the ambiguity associated with each of the bodies of thought. A consensus is however found between the theories on a number of issues. For example, each agree that there was a period of growth from the inter-war years until the 1970s when a crisis occurred within the capitalist system. Although each of the theories acknowledge the demise of the Fordist era, they differ in their perceptions of the succeeding conditions of economic activity. Each of the three theories are expounded briefly.

Flexible specialization theory

Flexible specialization is concerned with the restructuring of production through the adoption of a new technological paradigm (Piore and Sabel, 1984). The theorists propose that the market and technological conditions associated with Fordism no longer exist and that capitalists have embraced a new technological paradigm which forms the basis of the 'second industrial divide' (Piore and Sabel, 1984) as it overcomes the structural changes in the market and leads to a reorganisation in production. The technological paradigm of mass production is therefore being replaced by flexible specialization. Instead of using dedicated technology, new microtechnology has been adopted which is capable of performing a range of tasks and enables firms to enter new and specialized market niches.

The Flexible Specialization school highlights that the mass market for standardized goods is saturated and rising levels of disposable income have led to the fragmentation of demand. Yet, the movement into the second industrial divide has only been facilitated through the development of flexible technologies, such as Computer Aided Manufacturing (CAM), Numerically Controlled machine tools (CNCs), Computer Aided Design (CAD), Computer Integrated Manufacturing (CIM), and Flexible Manufacturing Systems (FMS). Advanced flexible technology enables firms to compete on economies of scope, as contrasted to the traditional competitive advantage through economies of scale. Hence, smaller enterprises become viable through the use of flexible technology and their ability to manufacture customized products. Sabel (1989) proposes that the emerging flexible specialization will reverse the dominance of large firms in the economy and lead to the formation of *regional economies* consisting of a network of flexible firms where all firms are equal and no one firm dominates.

Through these newly formed industrial districts of small firms operating under the principles of flexible specialization, a reorganization of the labour process will occur, as characterized by the multi-skilling of workers. Workers will be taught the skills necessary for using flexible machinery to its maximum potential and switching quickly from the production of one product to another. In addition the traditional antagonism between workers and management will be overcome within the new system with the basis of worker-management relations being 'predicated on collaboration' (Piore and Sabel, 1984:278).

Despite widespread criticisms of the flexible specialization approach, most notably the lack of empirical evidence in support of the theory it has also been attributed significant attention. Nielsen (1991:121) attributes this interest to 'the clear message, the suggestive line of argument and the optimistic vision [of the theory].' Yet Nielsen also highlights that the theory 'also provides a rationale of the much wanted new 'offensive' strategy in countries where the power of the left is less defensively rooted than in Britain. However, the dangers are obvious if the chosen guide to the future industrial world is overly optimistic and ignores important preconditions and side-effects.'

The neo-Schumpeterian theory

The neo-Schumpeterian school of thought, or alternatively the techno-economic paradigm approach perceives technology as central to economic development. Herein, the radical innovations of information and communications technology represent a movement into a techno-economic paradigm, as characterized by flexibility.

The neo-Schumpeterian school perceives five periods of economic activity as manifest through the 'Kondratieff waves'. Each wave is implemented through the development of a cluster of technological innovations, termed a techno-economic paradigm. The transition between each of the waves is characterized by a period of structural crisis in the economy. To overcome these periods of instability, socio-institutional structures must realign with the new techno-economic paradigm, and it is only when this occurs that growth may once again be initiated. While some (Hall, 1985) see growth occurring in response to the clustering of innovations both sectorally and geographically, others (Freeman, 1987; Rothwell, 1982; Marshall, 1987) have attempted to integrate innovation into the wider framework of social forces, including R&D expenditure, availability of technology and venture capital and the extent to which local government policy is conducive to technological change. While technological change remains central to the introduction of a new wave, recent neo-Schumpeterian writings have stressed the importance of social and political forces in determining the geography of new innovations. For example, Marshall (1987:228) argues; 'The succession of leading industrial regions have provided the basis for regional social and political movements which have, at critical moments, contested and to varying degrees determined the ensuing course of national and sometimes international economic development'.

Like the flexible specialization theory, again the new paradigm is perceived as establishing a solution to the rigidities of the post war paradigm of mass production. Technology within the new paradigm eliminates the diseconomies of scale and the rigidity of the mass production assembly line through the adoption of flexible manufacturing systems. Small firms become economically viable within the new regime through the attainment of economies of scope and their ability to manufacture customized products in a manner which was previously not possible.

The neo-Schumpeterian theory, like that of Flexible Specialization has also been widely criticized - most notably by the Regulationists - for its technological determinism. For example, each new wave of economic development is initiated by technological developments with socio-institutional adjustments being secondary to the development process. In no instances are socio-institutional factors identified as the primary agent in generating a new wave of economic activity. In contrast Regulation theory emphasizes both technological and socio-institutional factors, lending substantially greater weight to the latter than found within the Flexible Specialization theory or that of the neo-Schumpeterian school.

Regulation theory

During the early 1970s when the international markets were in a period of disorder and instability following the oil crisis, a theory to explain these changes emerged from the French regulation school. Some of the most important contributions to this theory have come from Aglietta (1979), de Vroey (1984), Boyer and Mistral (1983), Boyer (1986), Lipietz (1986, 1987) and Leborgne and Lipietz (1988). Since this period however, widespread analysis of the basic concepts has led to a number of hybrid theories which blur the boundaries between regulation, flexible specialization and neo-Schumpeterian interpretations of structural change.

The regulationists argue that economic development under capitalism is inherently crisis prone. Yet, between periods of crisis, relative stability exists with the prevailing social relations and institutional forms and structures complementing each other. During these non-crisis periods, growth occurs and this proceeds until a non-correspondence between the technological and socio-institutional factors emerge, at which point a further crisis is initiated. The theory therefore seeks to ascertain the regulating mechanisms associated with this movement from one period of growth to another.

The Regulation school is similar to that of the Flexible Specialization and neo-Schumpeterian schools, in that it also acknowledges a crisis in the capitalist system - dominated by Fordist production practices and Keynesian politics - around the early 1970s. Regulation theory deviates from the other two approaches in terms of the causes of this crisis and indeed the characteristics of the succeeding era. Despite convergence on theoretical issues between the different schools of thought, substantial differences remain. For example, the Regulationists argue that one of the causes of the structural crisis was that the consumption pattern had changed due to saturated mass markets, with consumers demanding 'a greater variety of use values' (Nielsen, 1991:127) which was contrary to that provided through standardization. Similarly, Piore and Sable (1984) identified the saturation of mass markets and the subsequent fragmentation of consumer demand as contributory to the crisis of Fordism. Yet, while the Flexible Specialization school highlights this factor as the primary cause of the crisis, it represents only one of the overall determinants of crisis within the Regulation school. In addition, while changing technology is perceived as being important within each of the schools of thought, the Regulationists emphasize socio-institutional factors to a much greater extent than the neo-Schumpeterians or the Flexible Specializationists, with these theorists often being accused of technological determinism. For example, 'in the regulation approach the emergence of new technologies is not seen as the decisive factor in the dynamics of the crisis. Neither is information technology seen as offering a ready-made way out. Technology might point at an exit from the crisis but the key can only be found through social processes.' (Nielsen 1991:128). It should be noted however that Aglietta (1979) attributed technological change and the reorganisation of production as vital in overcoming the crisis of mass production.

While regulation theorists present conflicting accounts of the emerging form of accumulation, this is in part due to the varying historical and geographical trajectories of development and conditions which being unique to certain areas, are fundamental in characterising the emerging nature of accumulation. However, like the Flexible Specialization and neo-Schumpeterian schools, the Regulation approach also acknowledges the emergence of a flexible form of accumulation. This flexibility pervades all aspects of the production process, not being exclusive to technology as in the introduction of technologically advanced flexible machinery, but includes the establishment of flexibility within the labour process (both internally within the firm, and externally within the labour force), the production processes, the state in terms of the governance structures and the prevailing ideology of the state.

Regulation theory seeks through the use of an evolving methodology and the expertise of other approaches - such as the French *economie des conventions* school and the US social structures of accumulation approach - to establish an 'institutionally-grounded political economy' (Peck and Tickell, 1992:2). Moulaert and Swyngedouw (1989:329) contend that Regulation theory is "well placed to capture the perceived contradictions of capitalism: it relates the general primacy of the capital valorization process to the concrete features of periods of sustained accumulation and subsequent crisis formation."

Regulation theory does not seek to resolve the crisis tendencies of capitalism, but rather, by acknowledging them, it seeks to explain the spatial phenomenon present. This is stressed by Aglietta (1979:37) in that the principal aim of Regulation theory is to 'provide a scientific interpretation of the appearance of novelty in the social realm. The theory must seek to answer the question as to whether a new phenomenon is a change in form of the determinant relationships or instead the emergence of relations which can, if they develop, transform the mode of production itself.'

While the flexible specialization theory hypothesises about emerging labour and production relations based on evidence in the Third Italy, Baden Wurtenburg and Silicon Valley, it fails to provide a holistic framework wherein changes in corporate flexibility are linked to regional trajectories of development. In contrast, Regulation theory presents a comprehensive framework of analysis wherein the perceived movement towards increased flexibility in the late 20th Century is encapsulated within an evolutionary process of economic and political change throughout the past 100 years.

Central to the Regulationist perspective of economic change is the notion of 'regimes of accumulation'. Boyer (1986; as quoted in Moulaert and Swyngedouw 1989:330) defines the regime of accumulation as; "the ensemble of regularities that assure a general and relatively coherent progression of the accumulation process. This coherent whole absorbs or temporarily delays the distortions and disequilibria that are born out of the accumulation process itself".

Further, Lipietz (1988:31 and 1986:19 as quoted in Peck and Tickell, 1992:5), describes the regime of accumulation as; 'a way of dividing and systematically reallocating the social product. Over an extended period of time there is a certain convergence between the transformations of production (amount of capital invested, distribution amongst the branches, norms of production) and transformations in the conditions of final consumption (habits of consumption of wage earners and other social groups, collective expenditures...)'.

Boyer (1986) argues that there are five essential components to the accumulation system, namely;

1. The organisation of production and the worker's relationship to the means of production, including the spatial manifestation of labour and technical relations;
2. A conducive form of market and sector organisation which facilitates the reproduction of the accumulation of surplus value;
3. The distribution of produced value to assure the dynamic reproduction of different classes and different social groups and therefore of the mode of production, i.e. proletariat and capitalist classes;
4. The composition of demand;
5. The composition of a social and spatial division of labour which ensures the progression of the capitalist system of production.

These conditions suggest that there is a direct relationship between capitalism and other modes of production (Lipietz, 1986). Peck and Tickell (1991:5) clearly demonstrate this relationship by showing the existence of two elements to the regime of accumulation, namely, the 'accumulation system' and the 'mode of social regulation'. The accumulation system is that part of the regime which ensures that the individual decisions of capitalists to invest, are met by demand for their products, the valorization of which is realized through the market mechanism, - usually over a number of decades. In addition, the mode of social regulation (or 'mode of regulation'), as formulated by Lipietz (1986) refers to the habits, customs, norms and laws which ensure the unity and continued reproduction of the accumulation system.

For every accumulation system there will exist a mode of regulation, however, this coupling is unique in that it is a result of the historical factors present within the environment. There is not therefore a standardized mode of regulation which is applicable to all accumulation systems. Yet, if the accumulation system and mode of social regulation complement each other, then stabilisation will occur, although this will be in the short-run, as the inherent contradictions of the capitalist system will lead to a further crisis.

Lipietz (1987:15) succinctly points out that the mode of social regulation is not the inferior partner of the coupling and any determination of it by the accumulation system should be avoided. Instead, the resultant accumulation systems and the modes of social regulation are 'chance discoveries made in the course of human struggles, and if they are for a while successful it is only because they are able to ensure a certain regularity and a certain performance in social reproduction...however...the history of capitalism is full of experiments which led nowhere, aborted revolutions, abandoned prototypes and all sorts of monstrosities.' The predominant regimes of accumulation in the 20th Century, until the present form of accumulation, and their associated accumulation systems and modes of social regulation are outlined in Appendix 1.

The present regime of accumulation directly attacks the rigidities of Fordism, bringing with it a flexibility of the labour structure; the production process; space; the state; and the overall ideology. Table 1.1 concisely summarizes these changes.

The flexible production and circulation system incorporates a number of changes including the introduction of both new techniques of production and new products. There have been two types of technological advance, namely, the emergence of microprocessor technology which has allowed firms to organize their production in a more flexible manner, and secondly the introduction of microprocessor-based innovations which have facilitated rising productivity and output. From the Computer Numerically Controlled tools which existed, flexible manufacturing systems have emerged which have led to different functions within the production process being integrated.

Table 1.1
Fordist versus flexible regimes of accumulation

Fordist	*Flexible*
The production process	
Based on economies of scale	Based on economies of scope
Mass production of homogeneous goods	Small batch production of a variety of product types
Large buffer stocks and inventory	No stocks
Testing quality ex-post	Quality control part of process
Rejects concealed in buffer stocks	Immediate rejection of defective parts
Loss of production time because of long set-times, defective parts etc.	Reduction of lost time; diminishing the porosity of the working day
Resource driven	Demand driven
Vertical and Horizontal integration	(Quasi) vertical integration or disintegration
Cost reductions through wage control	Learning-by-doing integrated in long-term planning
Labour	
Single task performance by worker	Multiple tasks
Payment per rate	Personal payment (detailed bonus scheme)
Bureaucratic labour hierarchy	Individualized promotion schemes
High degree of job specialization	Elimination of job demarcation
No or little on-the-job training	Continuous on-the-job training
Vertical labour organization and internal labour market segmentation	More horizontal labour organization for core workers
No learning experience	On-the-job learning
Emphasis on diminishing worker's responsibility	Emphasis on worker's co-responsibility
No job security	No job security and poor labour conditions for temporary workers; increased informal activities
Space	
Functional spatial hierarchy	Spatial clustering and agglomeration
Spatial division of labour	Spatial integration or division of labour
Homogenisation of regional labour markets	Labour-market diversification
Worldwide sourcing of components and subcontractors	Spatial proximity of vertically quasi-integrated firms
Organization of the space of consumption through suburbanization	Organization of the space of consumption through urban concentration
Selective socio-spatial integration	Polarization of the social use of urban space
State	
Collective bargaining	Division or individualization: local or firm based negotiations
Socialization of welfare	Privatization of collective needs and social security
International stability through multilateral agreements	International destabilization; increased geopolitical tensions
Centralization	Decentralization and sharpened inter-regional or inter-city conflicts
Subsidy state or city	Entrepreneurial state or city

Table 1.1 (cont'd)

Fordist	*Flexible*
Indirect intervention in markets through income	Extensive direct state intervention in markets and price policies through procurement
National regional policies	Territorial regional policies
Firm financed Research and Development	State financed Research and Development
Industry led innovation	State-led innovation
<u>Ideology</u>	
Mass consumption of consumer durables	Individualized consumption
Modernism	Postmodernism
Totality or structural reform	Specificity or adaptation
Socialization	Individualization; the spectacle society

Source: Albrechts and Swyngedouw 1989 as reproduced in Moulaert and Swygedouw 1989:336

To Computer Aided Design and Computer Aided Manufacturing, Flexible Manufacturing Systems (FMS) may be combined which enables production to be integrated. This then reduces handling time throughout the production process and allows the system to be monitored easily. It should be noted that there is limited evidence of FMS being used in the industrial world, with this being attributed to the high costs involved in introducing it and the problems of integration which it entails. Further, the inability to apply Computer Integrated Manufacturing (CIM and FMS) across all industrial sectors stresses the importance of avoiding technological determinism when examining periods of accumulation.

Many countries were unable to produce goods efficiently under the Fordist regime, often because their markets were too small to facilitate mass consumption, therefore new market niches had to be captured, which frequently meant the production of a greater variety of specialized goods, in smaller batches than that of mass production, to meet individual tastes and preferences. This was found throughout Europe and especially Japan where cars were redesigned to meet new demands with respect to efficiency, style and colour etc.

Within the Japanese automobile industry, the introduction of flexible automation reduced the overall production time and enabled a higher turnover of capital. Dertouzos et al (1989:19) point out that the Japanese could organize die changes in five minutes, while the same task within an American company - still rooted in the principles of Fordism - took between eight and 24 hours to complete. Additionally, the notion of quality control being a built-in component of the production process, radically reduced the wastage of resources, both in terms of raw materials and human time, hence furthering the search for absolute surplus-value.

Stocks were also reduced through the close supervision of production runs and this reduced the total value of constant capital. Subcontracting and vertical disintegration were also intensified in this new regime, which further allowed an increase in the circulation speed of capital. Yet, one of the inherent difficulties of this system was that it demanded the introduction of new products and innovations, an element which exerted great stress upon the Research and Development (R&D) departments of many firms. Moulaert and Swyngedouw (1989:338) believe that, for the first time in the history of capitalism, it [was] possible to combine flexibility with high productivity, or…to combine technologically harnessed labour with product and process diversification.'

These changes have demanded a revolution in the organisation of production, as the hierarchical control and management of corporations is restructured, along with planning and industrial location, as agglomerations or 'regional economies' (Sabel, 1989) are formed on the basis of 'competition and cooperation' (Best, 1990). Yet the introduction of advanced telecommunication technology has radically altered the face of managerial information flows, creating a system whereby speed and flexibility are almost synonymous.

The second aspect, relating to a new flexibility in the spatial integration of firms refers to the integration of all aspects of the production process which necessitate the close monitoring of the whole process. This leads to spatial clustering and agglomeration to facilitate the easy movement of ideas throughout the system. Further, instead of the world-wide sourcing of components and subcontractors, characteristic of the Fordist regime, the new flexibility results in the spatial proximity of vertically quasi-integrated firms.

The third aspect, namely that of the state, is of vital importance in the adoption of a new flexibility, as it seeks to act through an effective legislative framework, to overcome the tendencies towards overaccumulation and hence prevent a crisis within the economic structure. It is essential that the state acts in providing markets for the goods produced, yet this will by necessity involve a movement from the Keynesian 'welfare' state to that of an entrepreneurial state, as the distribution of social demand is altered throughout society. The establishment of an entrepreneurial state will allow the deregulation and decentralization of the trade unions and thus permit the 'liberalisation of private capital initiatives in regional or local '*lieux de valorisation*' (Moulaert and Swyngedouw 1989:339).'

It has been argued that the formation of such a policy framework will further polarize the unemployed and those unable to adapt to the new capitalist accumulation process. The policy actions which would lead to growing inequality include, (Moulaert and Swyngedouw, 1989:339).

1. Deregulation of social security laws and corresponding administrative practices;
2. Acceptance of a quasi-informal economy;
3. State support only to profitable projects (through incentives such as tax cuts and enterprise zones);
4. Decentralization of state powers to intervene in the valorization of capital to regional and local levels of decision making.

Hence in the formation of government policy, emphasis has moved away from dealing with consumption, to the control of production, as epitomized by the fostering of an 'enterprise culture' within the United Kingdom.

The fourth factor, namely the overriding ideology as encompassed through flexibility involves the emergence of a new urban life style which is then translated to new consumption patterns. Historically urban areas have been associated with the presence of the poor and working-class sectors of the population, however, within the age of *postmodernity* (Harvey, 1989), the social composition of these areas is changing, with a new class of urban dwellers being implanted in the built environment whose ideology dictates an individualism in consumption patterns and consequently the emergence of a *yuppie culture*. The new consumption patterns are the focus of much attention as this has demanded production to meet specific needs. Moulaert and Swyngedouw (1989) illustrate a number of examples where this new ideology has resulted in changes in the environment, namely, the inner harbour renovations in Baltimore, Boston; London and Liverpool; renovated industrial complexes transformed into shopping centres; the emergence of fashionable cultural centres - Brussels; and inner city renovations such as in Antwerp and Lille.

Yet, despite these changes, it is perhaps the emergence of a flexible labour force which has attracted the most attention from policy makers and academics alike. Within Regulation theory, changes within the labour process are viewed as vital to the continuance of capitalism. These changes pervade both the accumulation system and the mode of social regulation. While the Regulation theorists argue that the main crisis of Fordism has been the very success of the regime in reaching its maximum level of productivity, the reorganisation of labour and the restructuring of the production process, enables productivity gains to be attained once more. Technology is important from the regulation perspective in that it enables capitalists to intensify the use of labour, with this increasing the level of deskilling and increasing managerial control (a factor which is cited by Aglietta (1979) as crucial to the continuance of capitalism). Yet, more recently post-Fordist theorists have pointed to a re-skilling of the labour force and a blurring of the demarcation between production and white-collar workers (Jessop et al., 1988).

In contrast to the Regulation approach, the Flexible Specializationists and neo-Schumpeterians, while also pointing to the prominence of labour in the new regime, argue that it is technology which alone facilitates the movement into a period of flexibility. Further, it is assumed by the flexible specializationists that the increased diversity within the production process demands flexible technology, and this demands that the flexible worker is 'a kind of multi-skilled craftsman' (Wood, 1989:15).

The growing importance of small firms[2] in regional development

A review of changes in the number of small firms in France, Germany, Italy, Japan, the United Kingdom and the United States concluded that 'there has been a recent increase in the share of total employment in small enterprises and establishments... While the magnitude of the increase varies considerably from country to country and across sectors, its significance rests primarily on the fact that it signifies the reversal of a substantial downward trend in the employment shares of small units that had prevailed for many decades.' (Sengenberger et al., 1990 as quoted in Storey 1994:28). OECD data also confirms this trend and demonstrates that there has been a general increase in self-employment over the 1979 to 1991 period[3] in most OECD countries, with approximately one in eight workers in the EU being self-employed.

It is believed that a paradigmatic shift has occurred in the attitude towards small firms since the 1960s. For example, 'during the Afghan and Iranian crises President Carter canceled all speaking engagements except to 2000 entrepreneurs attending a White House Conference on Small Businesses. This is an indication of the esteem in which the small firm and its owner are now held by the White House' (Reilly, 1980:60 as quoted in Storey, 1982:5). Over the 1970 to 1990 period attitudes towards small firms changed dramatically and subsequently, Governments began to view them as vehicles of innovation and change within the economy. Indeed within the United Kingdom, small firms were publicly perceived as 'the engine of new enterprise in the 1990s' (*The Times,* January 1st, 1992:2).

The change in government policy was induced by increased academic research into the role of small firms. As the merits of small firms were expounded, governments sought to develop policy which harnessed the potential of small firms. Small firms were no longer seen as second-class establishments, but as a dynamic sector of the economy which could make significant contributions to regional growth. Government perception of small firms was strongly influenced by the work of David Birch (1979) and others pertaining to the potential employment effects of small firms.

Within the United Kingdom, the 1970s witnessed increased interest in small firms from successive governments. Anthony Grant (12, June 1972), the then

Under-Secretary of State with responsibility for small firms, said that he was 'determined that small firms should be allowed to flourish and thrive in the freest possible environment, unhampered by unnecessary restrictions and unintentional discrimination'.

The Labour Government of 1974 reinforced the ideals of the previous Conservative Government, regarding small firms as being 'important in regional terms as a seed-bed of regional growth and a source of diversification and balance in the industrial structure' (Eric Heffer, 21, April, 1978, as quoted in Beesley and Wilson, 1982:184).

The desire was so strong to stimulate the presence of small firms within the economy, that in September, 1977, Lord Lever and Robert Cryer were commissioned to make a study of the problems of small firms and to recommend plausible solutions. Previously, on the 5 January, 1977, the Government led by the Rt Hon James Callaghan MP PM, appointed a committee chaired by Harold Wilson, to inquire into the 'role and functioning, at home and abroad, of financial institutions in the United Kingdom, and their value to the economy.' (Wilson Report, 1979:iii). Beesley and Wilson (1982:185) state that 'the Conservative Government of 1979 pursued the trend towards greater intervention in the affairs of small business, started by the previous Labour Government,' demonstrating this by a quotation from David Mitchell the Under-Secretary of State, responsible for small businesses, 30, July 1980, that 'we have far too few such businesses. We need many more. The balance between the incentives to start a business, the hurdles which those who start and the burdens they have to carry has been tipped so far that the logical person has not felt it worthwhile to start a business.' Mitchell then proceeded to explain that this tipping of the balance could be counteracted, by identifying and dismantling the burdens, to allow a greater number of potential entrepreneurs to start a business; increase the incentives for the formation of businesses; and finally to review the problems of finance to those firms already in operation (Beesley and Wilson, 1982:186).

The incentives which were offered to small firms within the UK consisted of three types namely;

1. Those measures necessary in the advancement of economic and social policy, including increased exports, rural development, regional assistance and technological change.
2. Non-financial aid such as training and advisory services.
3. Fiscal incentives to encourage the establishment of small businesses, increase the supply of venture capital, and provide financing for a loan guarantee scheme.

Traditional regional development theory proposes that the success or failure of firms is intricately linked to the performance of the regional economy. Core areas offer comparative advantages to firms through agglomeration economies, a large skilled external labour market, infrastructure and technologically advanced service and production support activities. In contrast, peripheral regions are often characterized by an industrial structure epitomized by declining traditional industry and a low level of skills in the external labour market. Spatial determinism, or as McDermott and Taylor (1982) describe it, *spatial fetishism*, therefore exists with business performance being dictated by the local or regional environment.

The flexible specialization theorists point to the emergence of conglomerations of craft based federated enterprises (Piore and Sabel, 1984) or the agglomerations of customized producers (Scott and Storper, 1988) as enabling small firms to overcome regional disadvantages. Spatial propinquity eases small firm access to inputs and outputs and thereby reduces the costs of production while simultaneously improving the overall efficiency of the firm (Walker, 1985). Within this conceptualization, the formation of clusters is central to the improved comparative performance of small firms in peripheral regions relative to core regions.

Yet, other research (Vaessen and Keeble, 1995) in examining the performance of small and medium enterprises (SME's) throughout Great Britain demonstrated that firms within peripheral areas, such as Wales, Scotland and Northern England were able to grow 'in environments which lack the range and scale of facilities and agglomeration advantages of the South East' (Vaessen and Keeble, 1995:24). Agglomeration economies were not central to the success or failure of small firms, but instead small firms overcame the constraints of the regional environment through pro-active competitive business strategies. The remainder of this section summarizes the main aspects of small firm success and failure, and the perceived ability of small firms to act as employment generators within regional economies.

Small firm success and failure

It has been estimated that 40 out of every 100 new firms will cease trading within the first three years of business (Storey and Johnson, 1987). In 1991, 47,000 newly established businesses collapsed in Great Britain, averaging 199 each working day. Approximately 10 per cent of the 460,000 new registrations for VAT during 1991 resulted in failures (*The Times,* 1.1.92). The failure rate of businesses was not uniform across the United Kingdom, with business failures in Northern Ireland increasing by 35 per cent from 1990 as compared to an increase of 80 per cent within the East Midlands and 77 per cent within Scotland, over the same period.

In many cases, business failure can act as a learning experience. Often small firms become so financially adrift and caught up within the web of operational complexities that the only way out is by bankruptcy or the liquidation of assets. However, the entrepreneur will have the opportunity to begin afresh, having used the past failures and mistakes as a learning process, thus increasing the possibility of success in the next venture. Therefore 'there must inevitably be failures among new firms, particularly as the number and range of people drawn into running their own business is increased...however, failure rates among new firms are too high' (Smallbone, 1988:34). Particularly within peripheral regions the entrepreneur must aim to maximise his opportunities while simultaneously ensuring that all constraints are minimized. (Joyce et al., 1988).

Coyne and Wright (1986:287) outline the two determinants of small firm success. They state that 'it is often noted that because small firms operate in markets which may more closely approach the economist's concept of a competitive market, they are forced to operate efficiently in order to survive. Thus efficiency and competition are seen as vital elements.' Shimizu, in his study of large Japanese firms, claims that the strategy of competitive advantage is a key strength in the success of a company. Shimizu (1990:2) outlines that the strategy of comparative advantage refers to 'a strategy adopted by a company (firm) aiming at its long-term existence and development in trying to set its business position at a higher level than other firms... involves deciding the firm's product domain according to changes in business environment, as well as maintaining and developing the company's strength which supports the product domain.'

A strength is regarded as a feature which is almost unique and therefore cannot be imitated, the most obvious being that of human strengths. Hirashi Yamauchi, president of Nintendo Co. Ltd believes that human or 'soft' characteristics are more difficult to copy than material or 'hard' ones. Thus the production technology is harder to imitate than product technology, hence providing Nintendo with a competitive edge, being manifest through a rapid growth in sales and profit (Shimizu, 1990:3).

The Japanese are not alone in this belief of a competitive advantage allowing successful production, indeed Porter (1980:35) stresses three basic strengths which are requisite to obtain a competitive advantage, these being 'cost leadership, differentiation and concentration,' with the most important being differentiation.

Many researchers believe that a high failure rate amongst small firms is because they enter markets already saturated with particular products. For example Kelly and Davis (1972:60 as quoted in Johnson and Darnell, 1976:13), state that; 'Most new businesses setting up in the manufacturing sector do not have anything new to offer. They do not generally rely on process or product innovations and have little administrative skill, and it is hardly surprising that the average life expectancy is low. The new small firm tends to enter industries with low capital, technological and managerial requirements, and has consequently to contend with fierce competition.' It is therefore the industries with the lowest barriers to entry where the greatest concentration of new small firms are found.

Schumpeter (1961) proposed that new products are needed by firms to allow exploitation within unsaturated and specialized market niches, as well as to fight against competition and technological change. Therefore in the present competitive market, firms need to have 'an innovative capacity for creating, developing or adapting new products' (Adams, 1982:67). Adams accentuates Schumpeter's belief that if innovation is lacking at national level the society will deteriorate into a 'centralized socialist bureaucracy, much on the same lines as Ancient Egypt (ibid.).' Innovation is therefore seen as the key to competitive advantage within the small firm sector.

Innovation within the small firm is defined as 'both incremental or evolutionary type developments involving modifications of existing company product and also radical innovation involving a technological breakthrough' (Adams 1982:68). Innovative activity will reduce the dependency of small firms upon large firms as evident through the fragmentation strategies of devolvement and disintegration (Shutt and Whittington, 1987). Coyne and Wright (1986) believe that small firms based in the technological sector provide the best example of innovation in Britain. The State of Small Business (1983) reported that small businesses were producing 2.5 times as many innovations per employee as large firms and bringing innovations to the market more quickly, at a rate of 2.22 years as opposed to 3.05 years for large firms. Therefore if this is happening as suggested by the empirical evidence, why are small firms not more successful in their contribution to gross domestic product?

Impediments exist which dampen the innovation process within small firms. There are grouped into three categories;

1. To be innovative, it is necessary for a firm to find highly skilled professionals, such as engineers and technicians who may deal with the technical aspects of research and development. Small firms, especially in their early phases, are constrained financially and this therefore limits the degree of professional assistance they can acquire. Thus a vicious circle is created, whereby low financial resources means no professional help can be obtained, and hence no product innovations are possible, subsequently the operation within an already saturated market leads to low returns, and the circle is further accentuated.
2. Because of the size of small firms, they do not obtain the economies of scale that are achieved by large firms. Therefore, problems exist with information flows, financial support and networks of assistance. Consequently information as to the latest research developments and market opportunities does not reach the small firm quickly enough, often resulting in a wastage of resources as money and time are invested in unproductive activities.

3. Another form of disadvantage due to size is the management structure. Usually the small firm consists of an owner-manager with limited professional training in aspects such as sales, production and marketing of the products. Where this is found, despite the extent of innovation that is occurring, if the correct marketing does not occur, the potential market will not be exhaustively exploited and thus, total revenue will be diminished.

Smallbone (1988:45) concludes that 'demand deficiency is the most serious problem facing the majority of newly established firms.' While Storey (1985) proposes that advisory agencies are unable to assist with this problem, Smallbone suggests that small firm development agencies have a responsibility to discuss in detail the feasibility of markets for new firms, and this could reduce significantly the proportion of small business failures. Small business success within regional environments is therefore closely related to product differentiation through innovation which enables the small firm to gain competitive advantages within specialist niche markets, and process innovation which increases the overall efficiency and competitiveness of the small firm relative to its larger rivals.

Small firms and their contribution to employment

In 1979 David Birch produced the results of his seminal study on changes in urban employment throughout the USA. Birch's findings were summarized as follows (Giaoutzi 1988:6):

1. Gross job loss through closure and contraction was about 8 per cent per year.
2. About 50 per cent of the gross job gains were created through expansions of existing companies and about 50 per cent through new openings.
3. About 50 per cent of gross jobs created by openings were produced by independent, free-standing entrepreneurs and 50 per cent by multi-plant corporations (in-moves).
4. Firms with less than 20 employees generated 66 per cent of net new jobs in the United States.

Of all of Birch's results it was however the last result pertaining to the employment generated by very small firms which received most attention. The US Small Business Administration (SBA) directed the Brookings Institute to use the same data source as Birch had used - namely the Dun and Bradstreet data base - for the period 1978 to 1980[4] and establish the validity or inaccuracy of Birch's findings. The Brookings Institute research, as co-authored by Armington and Odle (1982) discovered that for this two year period, firms with under 100 employees generated 38 per cent of the employment, a figure approximately equal to their share of the labour force (Giaoutzi, 1988:7).

As a result of the conflicting evidence as published by Birch (1979) and Armington and Odle (1982), the SBA then asked Birch to re-calculate the extent of job generation which was attributable to the firms with under 100 employees, for the same period as Armington and Odle had investigated, namely 1978-80. Birch however produced a job generation figure of 70 per cent, that is 32 per cent higher than that found by the Brookings Institute.

Due to the conflicting results similar analyses were also performed both within the United States and the United Kingdom. Storey and Johnson (1987) concluded that generally the results as produced by Birch, over-estimated the extent of job generation which was attributable to small firms, while simultaneously that by Armington and Odle, underestimated the extent. Thus, Storey and Johnson (1987) concluded that a more accurate figure was 50 per cent of new jobs being created as a result of firms with less than 100 employees, which represented 38 per cent of the work-force. Subsequently, Storey and Johnson's findings served to affirm the findings of Birch (although to a lesser extent) that the small firm sector made a greater contribution to employment generation than larger firms.

A number of further studies were carried out throughout Europe, investigating the effect of small firms upon employment growth. Within the United Kingdom, Gallagher and Stewart (1984) discovered that of all the new jobs which were created within the private sector during the years from 1971 to 1981, 31 per cent were attributable to the small firm sector, which represented only 13 per cent of the workforce in 1971. Subsequently, such findings were welcomed by Government officials who were keen to uncover answers to the burgeoning problems of unemployment. In their enthusiasm however other aspects of Gallagher and Stewart's work were ignored, namely their declaration that; 'Our results must be interpreted with some caution...No one size of firms is responsible for the majority of job creation in the UK. Rather, all size sectors make a significant contribution. The results of this study do not follow the pattern of either the USA that most jobs stem from small firms or... that few jobs stem from small firms' (Gallagher and Stewart, 1984, as quoted in Storey and Johnson, 1987:30).

The work of Birch, Armington and Odle and Gallagher and Stewart was seized upon by policy makers in their enthusiasm to find a quick and palatable solution to the problems of unemployment. Giaoutzi (1988:2) concisely summarizes the situation in that he states; 'In recent years when the economies of most industrialized countries have shown various signs of stagnation and structural decline, a variety of studies have been undertaken which demonstrate clearly that small and medium-sized firms may be regarded as generators of new growth, as primary sources of technological change and, via job creation, as one of the major factors in maintaining economic stability'.

Governments therefore have jumped onto the job creation band-wagon of small firm development policies, in an attempt to stabilize the mode of social regulation and generate overall social, economic and political stability within the economies. This is most aptly illustrated through the Governmental White Paper (Cmnd 9794, 1986 - as quoted in Storey and Johnson, 1987:3) whereby the primary aim of Government was to 'encourage the development of an enterprise economy. The way to reduce unemployment (was) through more businesses, more self-employment and greater wealth-creation all leading to more jobs...(with key aspects of the Department's work being) to promote enterprise and job creation in growth areas such as small firms, self-employment and tourism'. Irrespective of the actual job generating potential of small firms, the perceived panacea of small firms as catalysts of regional rejuvenation has been warmly embraced by policy makers since the 1970s.

The peripherality of Northern Ireland within the United Kingdom

While the overall economic performance of the UK has been described as being peripheralized relative to that of Western Europe, the incidence of peripherality and deprivation have been characterized by regional inequality. Northern Ireland has been one of the regions of the UK which has been under-performing with respect to GDP and other variables of economic prosperity. Borooah (1993:2) states that, 'This picture of relative economic decline in the UK becomes particularly dramatic when one narrows the focus to Northern Ireland: arguably, on every economic indicator Northern Ireland has performed more poorly than the UK as a whole and, indeed, relative to most regions of the UK.'

Borooah then outlines some of the economic indicators which demonstrate this inferior position. For example, as outlined by the Family Expenditure Survey (FES), in 1990 Northern Ireland was recognized as having the lowest standard of living within the UK; unemployment in Northern Ireland on February 1992 was 14.2 per cent while the comparative figure within the UK was 9.4 per cent; and manufacturing employment had fallen persistently since the post-war years when it peaked at 195,000 in 1950 comprising 36 per cent of the total Northern Ireland employment, to 104,000 in 1990 or 18 per cent of the total Northern Ireland employment level.

While the Northern Ireland economy has continued to under-perform relative to the UK over a number of decades, questions are still being asked with respect to the reasons for this. One economic reason which has been proposed relates to *peripherality* whereby the geographical positioning of Northern Ireland at the edge of major population concentrations and areas of high economic activity has rendered the region disadvantaged. For example geographical distancing from major markets may create costs for firms in both tangible and intangible forms. While peripherality is usually associated with increased costs of transportation, the significance of these additional costs in the overall performance of the firm has been widely debated with some concluding that the peripheral position of Northern Ireland does not result in significantly higher costs (Hitchens and Birnie, 1989).

Aside from additional costs of transportation, it is indisputable that other costs exist, albeit intangible and unquantifiable, due to the peripherality of Northern Ireland. For example, Clulow and Teague (1993) suggest that peripherality may manifest itself in a constrained local labour market. Within this situation, the number of alternative employment opportunities will be reduced and the fluidity of the external labour market diminished. Therefore, managers who encounter a constrained level of labour turnover 'will tend to invest less in training and skill development as employers gravitate towards cost rather than commitment orientated internal labour markets' (Clulow and Teague, 1993:108). Clulow and Teague (1993) also postulate that peripheralization of the Northern Ireland economy will result in higher transaction costs for businesses. For a market to operate efficiently it is vital that there is a free flow of information to minimize the risks of investment. Within Northern Ireland the problem exists whereby, due to geographical peripheralization combined with the need for buyers and sellers to undertake face-to-face communications and transactions, this culminates in escalated transaction costs. Clulow and Teague (1993:109) thus state that 'to reduce uncertainty, suppliers and customers will try to establish long-term relationships in which a given exchange is repeated many times. In many instances these long-term relations are normally based on established social relations. The social links involved are of many kinds - formal and informal associations, ties of culture in particular - but they are all influenced by history and tradition, usually in a national or local context.'

Disproportionately high transaction costs within the periphery relative to the core regions of economic activity are therefore one of the key factors in determining the market distribution of products. For example, Hitchens, Wagner and Birnie (1989) found that Northern Ireland firms distributed a larger proportion of their output within the local market area as compared to Matched-Plants in Germany. While one of the reasons which they proposed for this was the inferior finish and overall low quality of the products manufactured in Northern Ireland, a further covert explanation would relate to the higher incidence of transaction costs within Northern Ireland as compared to Germany thus discouraging the fostering of external and export markets by Northern Ireland managers.

Historically therefore, peripherality has been an influential factor in determining the level of economic activity within Northern Ireland. While it is not always possible to quantify the effects of peripherality, it is crucial that the presence and pervasiveness of such is remembered when analysing the performance of firms within the region. While the effects of peripherality within Northern Ireland, both spatially and economically, have been ascribed limited validity within the research literature, other theories have been granted greater recognition. These theories have looked beyond the perceived peripheralization of Northern Ireland and instead have attempted to identify the inherent causes of an inferior economic position within the world market. This is important because if the attributes of peripherality within the Northern Ireland economy are examined and policies or strategies thus developed to minimize the impact amongst the business community, then this in part accepts the peripheralized position while failing to search for ways to counteract the tendencies of inequality. For example, the existence of cores and peripheries may generate processes which are self perpetuating with growing inequality between the core and periphery. Cores may accentuate their dominance through processes of exploitation or exclusion on the periphery. Hence, through exploitation 'the absolute or relative poverty of peripheral and semi-peripheral states continually induces their rulers and subjects to participate in the world division of labor for marginal rewards that leave the bulk of the benefits to the rulers and subjects of core states. Processes of exclusion, in contrast, refer to the fact that the oligarchic wealth of core states provides their rulers and subjects with the means necessary to exclude or crowd out the rulers and subjects of peripheral and semi-peripheral states from the use and enjoyment of resources that are scarce or subject to congestion. The two processes are distinct but complementary' (Arrighi, 1990:17).

Finding the sources of inequality must therefore be identified and reversed if the counteracting tendencies of economic development within the peripheries are to be overcome. The most comprehensive attempts to obtain this are evident within the theories founded upon the concept of structural change. These include the neo-Schumpeterian, Flexible Specialization and Regulationist schools of thought. While each of these bodies are distinct, they all acknowledge the demise of the Fordist era of production and the evolution of a new system characterized by flexibility.

Conclusions

This chapter has highlighted that traditional explanations of industrial location and the subsequent economic prosperity of regions are being challenged. The product life cycle theory and the associated New International Division of Labour and urban hierarchy model, are being replaced by non-hierarchical models concerning regions' innovation systems. Within the new theories each region is ascribed a competitive advantage based upon its inherited characteristics, the behaviour of firms, the innovative milieu and the regional policy. It is therefore the combination of these factors which determine the technological trajectory of peripheries as independent from national and international hierarchies.

Central to this ability of regions to determine their own economic prosperity is the emergence of flexibility as a means of overcoming limits to productivity and serving increasingly diverse market demands. Despite differences between the Flexible Specialization, Neo-Schumpeterian and Regulationist schools concerning the cause and nature of the crisis of mass production, concensus is found on increased flexibility within the labour and production processes. Technological advances have enabled firms to attain competitive advantages based not only upon economies of scale, but also upon economies of scope. Organisational restructuring within large firms and the changing nature of labour markets has led to small firms occupying a growing proportion of total employment. Small firms are now perceived as engines of economic growth and have been warmly embraced by governments as generators of employment.

For peripheral regions such as Northern Ireland, the emerging theories of industrial location and structural change suggest that the vicious circles of marginalisation can be overcome. Traditional core-periphery relationships can be challenged as regions determine their own innovative milieu and subsequently their own prosperity. Chapter two examines the characteristics of the Northern Ireland economy and contrasts it with a region which is proclaimed to have overcome the rigidities of Fordism as characterized by flexibility, namely Massachusetts.

Notes

1 Oakey (1984: Chapter 4), presents a comprehensive overview of the advantages of agglomeration to firms in the development stage of product development.

2 Small firms are defined as those establishments which have a relatively small share of the market, are managed by owners rather than through a formulated management structure, are independent of larger companies or organisations and have 50 or less employees.

3 See also Acs, Audretsch and Evans (1991) who examine the self-employment rate in 23 OECD countries from the mid 1960s to the late 1980s.

4 Birch had studied the period 1969 to 1976.

2 Vicious and virtuous circles of regional development - The case of Northern Ireland and Massachusetts

Introduction

While both Northern Ireland and Massachusetts experienced buoyant economies during the post war years founded upon the traditional industries, the transformation of the industrial structure has followed diverse trajectories since the 1970s. While the Massachusetts economy has managed to overcome its periods of crisis Northern Ireland has grown more dependent upon transfer payments from Westminster and the featherbedding of the local economy. Within Northern Ireland, subvention from the UK treasury has grown steadily since the early 1970s, when it was approximately 15 per cent of Government spending, to 35 per cent of Government spending in the 1980s. It is suggested that an ever increasing public sector within Northern Ireland may 'squeeze out the exposed traded sector' (Bradley and Wright, 1993:34). Yet, the substantial trade deficit within Northern Ireland is not only a 'consequence of the external financing of the Northern budget deficit by means of the annual subvention from London, but may also be symptomatic of a lack of international competitiveness'. (ibid.)

The emergence of the New International Division of Labour stimulated the decline of the traditional industries within both Northern Ireland and Massachusetts. This was in response to increasing international competition, import substitution, capital flight and plant closures. While both regions were severely affected by the relocation of large traditional manufacturers, the increasing dependence upon inward investors within Northern Ireland led to large job losses in the late 1970s and early 1980s. Within Massachusetts however the relocation of traditional industry was perceived as a window of opportunity through which high technology firms could acquire premises which were large, cheap and conveniently located in terms of infrastructural requirements. The increasing role of the high technology sector was also closely dependent upon the research reputation and capabilities of the universities within the state, the ready supply of low-skilled, low wage entry-level workers, the specialist skills of engineers and scientists, the political leverage within the Pentagon which stimulated the disproportionate concentration of prime contracts within the state, and the relative lack of unionized labour within the high technology sector. Therefore, 'The fortuitous juxtaposition of a number of unplanned and probably unplannable factors converged in New England to create the seedbed for the new high technology industry' (Frankel and Howell, 1985:303).

In contrast to Massachusetts the Northern Ireland economy failed to restructure itself, and in the 1990s remains disproportionately biased towards those traditional industries which are in terminal decline throughout Europe. Within Massachusetts the disintegration of large firms (Shutt and Whittington, 1987) has led to the formation of agglomeration economies (Piore and Sabel, 1984) as epitomized by the high technology agglomeration along Route 128. Northern Ireland however has been unable to develop a technological strength in any industrial sector being constrained on the input-side, by the absence of high quality local suppliers and labour supplies (see Oakey, 1984).

This chapter presents the main changes in the economic profiles of Northern Ireland and Massachusetts focusing specifically upon the divergent paths which they have taken following the decline of the traditional industries. Changes in the industrial structure are also examined with further analysis being made of the relative position of the manufacturing sector in the changing industrial base. The composition of the manufacturing sector is also analysed with the inertia of the Northern Ireland manufacturing sector being contrasted with a dynamic manufacturing sector within Massachusetts. A measure of regional vitality over the 1973 to 1994 period is then presented through unemployment rates. The examination concludes that while both regions have experienced similar structural problems since the 1970s, the Northern Ireland economy has become entrenched in a vicious circle of development as characterized by an uncompetitive manufacturing base and an over-dependence upon income-sustaining transfers from both the UK Treasury and the EU. In contrast, continued restructuring and revitalisation of the Massachusetts economy in response to changing structural conditions has been facilitated through a virtuous circle of development based upon strategic alliances between universities, research institutions, private companies and government bodies.

Economic development in Northern Ireland

Between 1950 and 1958 the rate of growth of Gross Domestic Product (GDP) was similar in Northern Ireland to the rest of the United Kingdom, at 1.7 per cent per annum. Yet by 1960 GDP per capita was only 63 per cent of the United Kingdom level (Hitchens and Birnie, 1992:6). Not only was output in Northern Ireland considerably lower than in the rest of the United Kingdom, but the sale of Northern Ireland products was highly dependent upon the British market. By 1960, 61 per cent of goods manufactured in Northern Ireland were sold to the Great Britain market, with a further 30 per cent remaining within the Northern Ireland market (Harris, 1991:24).

The structural benefits of industry during the war years were rapidly turning into a handicap. Between 1950 and 1958 agricultural employment in Northern Ireland declined by 20 per cent and employment in the Northern Ireland textile industry fell by 30 per cent. These contractions in employment were further accentuated by low levels of investment in the economy. Low levels of investment during this period were attributed to; a low level of profitability; the general social instability of the region; and high transportation costs of both raw materials and the finished products as a result of the geographical isolation of the province (NIEC, 1983:3). The government therefore intervened in the market to encourage greater private sector investment.

By the 1950s, the war time dependence upon the textile industry (which had accounted for over half the manufacturing employment) had disappeared and this led to the structural decline of this industrial sector. The government therefore believed that the attraction of inward investment would be advantageous in both the short-term and the long-term as it would both increase employment and diversify the industrial base of the region. A range of schemes offering financial assistance to investors was established through the 1954 Capital Grants Scheme, which provided enticements, including such *carrots* as interest free loans, grants, tax relief, advance factory buildings, exemption from taxes and grants to aid the relocation of workers.

In 1966, grants were introduced to help reduce the initial high start-up costs of new investments within Northern Ireland. Between 1967 and 1970, employment grants accounted for £12.8 million or 6.9 per cent of total grants issued through Industrial Development Assistance (Harris, 1991). Additionally, the Government provided indirect stimuli to make firms locate or relocate in 'peripheral' or less desirable areas, in the form of Industrial Development Certificates, as well as establishing planning controls which restricted the ability of firms to expand, and therefore encouraged them to relocate in more peripheral regions of the United Kingdom.

Initially, the Government offered advance factory buildings, but this was soon expanded to include machinery, plant and building grants at a quarter of the total cost of the investment, through the Capital Grants Scheme of 1954. This was increased in 1959 with the objective of rejuvenating indigenous industry within the economy while simultaneously encouraging the substitution of capital for labour to increase productivity, profitability, efficiency and, most importantly, employment. This form of selective assistance was continued in 1966, in the form of Investment Grants, which provided 40 per cent - rising to 45 per cent in 1967/68 - of the total costs of machinery, plant, or buildings, yet in this case, the aim was not to rejuvenate existing companies, but rather to increase both the absolute level of investment and the productive capital stock, a factor which culminated in the relocation of industry on a national scale. [1]

Government policy since the 1960s has tried to attract industry to the region and counteract the structural problems of over-dependence upon shipbuilding and textiles, by diversifying the industrial base, while simultaneously reducing unemployment levels and providing a multiplier effect upon local industrial development. The success of Industrial Development policy in achieving these aims has been examined elsewhere (see NIEC, 1983b; Bull, 1984). The data suggests that inward investment has largely been concentrated within a few sectors, namely textiles, instrument and electrical engineering as well as vehicles and aircrafts, food, drink and tobacco, and clothing and footwear. Assistance through Industrial Development (ID), did not generate employment uniformly across the manufacturing sector, but rather, larger contributions were made within certain industries.

Although government policy continued to encourage inward investors during the 1970s, the consultants of the 1970 to 1975 Northern Ireland development plan stressed the need to increase industrial development through further grants. Yet, they acknowledged that, 'the full list of the various kinds of assistance already available in Northern Ireland undoubtedly conveys the impression that the area is already so heavily subsidized that any further assistance would be hard to justify', (Matthew, Wilson and Parkinson, 1970:44). The 1972 Industry Act therefore led to the introduction of selective financial assistance (SFA) as a means by which government could tailor financial assistance to the individual needs of investors.

The continuance of ID assistance in the 1970s reflected both the overall decline in the number of expansion projects in search of grant aid and the economic problems which many firms were facing during this period. Employment had been closely aligned with industrial policy within Northern Ireland. For example, employment within some industrial sectors in the 1960s and 1970s, most notably instrument and electrical engineering, textiles, clothing and footwear, and other manufacturing, was almost all in ID assisted firms (Harris et al, 1990:105). Harrison (1991:106) suggests that between 1959 and 1974 employment due to ID assistance rose by 3 per cent, to 7 per cent per year, whereas non-ID assisted employment continued to decline. Yet, this was to be a temporary phenomenon especially within the textiles and man-made fibres industry where 8,600 ID jobs were to be lost between 1979 and 1982, representing 61.3 per cent of the total ID employment lost during this period. The major reason for this collapse in the industry was the closure of major producers such as ICI, Courtaulds and British Enkalon within Northern Ireland.

Between 1973 and 1986, 60.5 per cent of the decline in total manufacturing was attributed to the closure of externally owned companies which had moved into the province between 1945 and 1973 largely under the auspices of ID assistance. In comparison, the Republic of Ireland and Leicestershire, although also experiencing a decline in total manufacturing employment during the years 1973 to 1986, of pre-1973 externally owned companies, were not as severely affected, losing employment of 41.9 per cent and 32.8 per cent respectively. Closer inspection of the data also suggests that Northern Ireland experienced a higher level of job loss over the same period through the contraction of pre-1973 externally owned companies, namely 55.7 per cent; as compared to the Republic of Ireland - 34.4 per cent which includes 'other' externally owned - and Leicestershire - 29.3 per cent (NIERC, 1989:35).

By 1980, the aggregate value of government grants and subsidies accounted for over 20 per cent of manufacturing GDP, yet by the end of the 1980s this had declined to nearer 10 per cent (Murshed et al, 1993). During the 1970s and 1980s the maintenance of jobs became much more significant than jobs promoted, as measured by government assistance (NI Annual Abstract, 1983; Harris, 1991) and during the 1980s between 80 per cent and 90 per cent of financial assistance went to established as opposed to new firms.

In the early 1990s there was a change in policy emphasis away from financial assistance based upon job creations, to the provision of financial assistance in improving the competitiveness of firms. By 1996 however, despite a reduction in the number of capital grants awarded by the small firm development agency, inward investors were still being offered financial assistance based upon jobs promoted.

The loses in manufacturing employment during the 1970s prompted a number of initiatives aimed at improving 'the general environment of Northern Ireland... The possibility of attracting industry depends not only upon the offer of subsidies or the existence of persuasion [but] ... By appropriate measures it should be possible to create conditions that will strengthen indigenous industry as well,' (Matthew, Wilson and Parkinson, 1970:13). These initatives related to investments in infrastructure and related services, the provision of which it was believed would create 40,000 new jobs (these expectations were regarded as optimistic, due to the negative externalities associated with political and social instability of Northern Ireland at this time). While the crux of the 1970-75 Northern Ireland development plan was the further attraction of inward investment, greater attention was now being directed towards the indigenous sector as a possible area of industrial expansion and growth. The Local Enterprise and Development Unit was therefore established in 1971 with responsibility for small indigenous firms while the Northern Ireland Finance

Corporation was formed with powers to allocate substantial amounts of funding at the direction of the government. The remit of the Local Enterprise and Development Unit and the Northern Ireland Finance Corporation are outlined briefly.

The Local Enterprise and Development Unit

The Local Enterprise and Development Unit (LEDU) was established in 1971 as a response to the claim that; 'the prospects of securing a satisfactory level of employment in manufacturing industry would be enhanced if more local initiatives were shown in the establishment of new businesses and the expansion of existing ones.' (Matthews, Wilson and Parkinson, 1970:81). By means of a grant from the Department of Commerce, LEDU was established with the following objectives (Harvey, 1983:87):

> To encourage employment in established small firms (that is <50 employees) through a greater exploitation of manufacturing and marketing opportunities along with a more effective use of resources, by providing selective assistance, factories and sites for the building of factories and acting in an advisory capacity as appropriate;
>
> To encourage the establishment of new enterprises by creating the right conditions or establishing groups of people at local or community level;
>
> To foster craft industries by taking measures to improve standards of quality and design and in marketing by collective promotion of people in the various craft disciplines.

Since 1971, LEDU's primary aim has been to create additional employment in the small firm sector in Northern Ireland. Since 1990, LEDU has aimed 'to strengthen the Northern Ireland economy by encouraging enterprise and stimulating improvements in the competitiveness of new and existing businesses within defined markets.' (LEDU, 1996:3). Between 1989 and 1994, employment created in surviving LEDU assisted firms was sufficient to significantly outweigh employment lost through closures. However, employment created among new firms starting up between 1989 and 1994 was the major source of employment gains.

The Northern Ireland finance corporation

The Northern Ireland Finance Corporation was formed in 1972 under the NIFC (NI) Order 1972, with the directive of assisting the economy of Northern

Ireland through aid to start-ups, growth firms and firms experiencing financial pressures. These processes it could achieve through the control of stock; provision of loans and by providing the guarantee for loans made by others. For each category of need a specific response was available. Firms faced with liquidity problems obtained assistance through the 'survival' category and in this case, the NIFC would provide equity capital to avoid a reduction in production or even closure. Growth firms which had developed to a point where they were unable to expand their production or increase investment further, were expected to apply for assistance through their banks. However, where firms required finance beyond what they were able to secure, the Northern Ireland Finance Corporation intervened, often in the form of long-term support.

The final category, through which assistance was granted was 'New Industries'. Herein, any of the above measures could be implemented to facilitate new firms becoming established within the province. This was perceived as, 'adding to the regional development package available' (Murie et al, 1974:122), and was the most significant of the three categories in achieving the objectives of the Corporation.

The assistance provided by the Corporation to the economy was achieved through the following methods (Harvey, 1983:83):

> Preserving the essential fabric of industry and commerce;
>
> Helping to establish an improved industrial structure upon which a sounder economy could be developed.

The Corporation only existed for four years being replaced under the 1976 Industries Development Order by the Northern Ireland Development Agency (NIDA) which had an expanded role of assistance and was accountable to Parliament. Although strict guidelines were laid down for the NIDA, accurate statistics on job promotions and maintenance are not provided by the Agency and this makes an evaluation of its success almost impossible. Yet, Harris (1991:84) states that the expenditure by NIDA between 1976 and 1982 amounted to £57.9 million in 1980 prices, with £4.3 million attributable to selective aid in response to directives from the Department Of Commerce; while £53.6 million went to the provision of equity and loans.

In September 1982 the NIDA, along with the Industrial Development Organisation which had provided financial assistance to ID projects, were joined together, with the new organisation being known as the Industrial Development Board (IDB). The IDB was concerned primarily with firms employing in excess of 50 employees, and as such complemented the work of LEDU. Harris (1991:85) illustrates that during the 1980s, the relative position of the two organisations - namely, LEDU and the IDB - altered radically, as manifest by a three-fold increase in the level of Government support to LEDU, while simultaneously, the level of support for the large firm sector, as incorporated within the IDB and its counterparts, fell considerably, namely by £48.82 million (at 1980 prices). Therefore a significant change in policy emphasis occurred during the late 1970s and early 1980s within Northern Ireland. This new direction in policy formation was directed more at encouraging the formation of new indigenous enterprises and the expansion of existing ones as opposed to attracting inward investors.

The changing emphasis of industrial policy in the 1980s

During the 1980s there was a growing awareness of the ineffectiveness of automatic grants and subsidies and instead selective assistance was adopted as a viable alternative. In 1985, 'Encouraging Enterprise - A Medium-Term Strategy for 1985 to 1990', was published, with particular emphasis being placed upon the development of an enterprise culture in the region. The document argued for assistance moving away from direct financial incentives to that of encouraging enterprise through fostering improvements in training and marketing etc. 'Encouraging Enterprise' was however criticized for being too general in its approach (Harris, 1991:88) and therefore, in July 1987 the Department of Economic Development (DED) published the 'Pathfinder' report[2] . This discussion document identified a number of key weaknesses within the local economy namely, 'a lack of enterprise, poor competitiveness, weak export performance, a large public sector, the use of public funds and effectiveness' (NIEC, 1990:7).

The Pathfinder report (DED, 1987) reiterated the emerging ideology within Great Britain and concluded that there were two objectives which the economic development policy within Northern Ireland should be founded upon, namely; '(a) To strengthen the Northern Ireland economy to provide self-sustaining employment; and (b) To support existing employment in Northern Ireland where it is cost-effective to do so.' (NIEC, 1990:7).

The Pathfinder report was however widely criticized for a number of reasons. The failures of the report included, (Clulow and Teague, 1993:69):

> Its disregard of the role of inward investment, which had played a major part in the employment generation between 1973 and 1986 (Gudgin et al, 1989);
>
> Its assertion that Northern Ireland could become a self-sustaining economy was perceived as being unrealistic and only served to represent a clarion call, without providing the means through which this could be attained;

Its disregard for the institutional framework and the other wider environmental conditions. That is, the report ignored that presence within the Northern Ireland economy of a 'disfunction between the regulatory or institutional framework and underlying market conditions, thereby creating economic malaise' (Clulow and Teague, 1993:69), and therefore by trying to impose an economic policy model based upon the minimization of institutional presence within the market acted in an oblivious manner, to the potential realization of their goals.

Economic policy was therefore reviewed and the IDB published a new document in May 1990, entitled 'Northern Ireland Competing in the 1990s: The Key to Growth'. This document attempted to overcome the inherent limitations of the Pathfinder report and while still focusing upon improving the overall competitiveness of the Northern Ireland economy, stressed the significance of inward investment and the proactive role of the private sector. The policy remit of the agencies which operate under the guidance of the DED - that is, the IDB, LEDU and the Training & Employment Agency (T&EA) - was radically altered with the movement away from direct forms of 'hard' assistance in the form of subsidies and grants, to that of 'soft' assistance, such as the promotion of training, R&D, quality, design and marketing. This change in policy emphasis was epitomized by the DED statement (1990:17 as quoted in Harris, 1993:82), that the role of the government would be to: 'Increasingly target its support for industry on areas such as training, research and development, quality and design rather than on capital investment...the government wishes to move to a position where financial assistance will become conditional upon evidence that the business applying for assistance is actively taking the right steps... Financial assistance will no longer be related solely to the prospects of additional employment...but rather to greater efficiency out of which additional long-term employment will come.'

Within LEDU, net job creation still remained a fundamental goal of the agency under the new strategy (Buckland, 1993), yet the manner in which this was to be achieved changed. Instead of providing grants on the condition that a firm increased its employment levels, the 'inhibitors and accelerators of growth' became the primary focus of the new strategy. LEDU foresaw that a transformation in the attitudes of firms towards the agency was essential, from being a source of available finance, towards that of assisting in the development and growth of the firms. They therefore attempted to develop a 'tripartite relationship' (Buckland, 1993:13) between the firms, LEDU and consultants in the analysis of company needs and in the prescriptive forms of assistance as determined by LEDU.

While the overall objective of the new policy is to improve the competitive position of firms, uncertainty abounds as to how this can be measured and especially how the contribution of the agencies can be assessed. Research has however demonstrated that LEDU assisted small firms displayed higher survival rates and growth rates than non-assisted firms (Scott, 1993). What is clear is that a decline in the proportion of inward investors relative to total establishments has been accompanied by a growth in the small indigenous sector. Yet, as analysis shows in the latter part of this chapter, manufacturing activity remains concentrated within the low wealth creating traditional industrial sectors such as food, drink and tobacco, and textiles. While it is argued that 'if policies to increase the competitiveness of small firms can be extended and made more effective, this sector provides a substantial base for future success', unless competitiveness can be nurtured within high price range products and science based industries which have the potential to introduce product innovations, then it is unlikely that this will improve the competitiveness of the Northern Ireland economy.

Economic development in Massachusetts

Prior to the US Revolution the majority of colonists were farmers with a small proportion being employed in small-scale manufacturing. Ferguson and Ladd (1988) point out that manufacturing was significant within the state since the 17th Century when the colony granted patents to the state for the manufacture of agricultural tools. Subsequently, the skills attained from this form of manufacture provided the foundation upon which the machine tool industry of the region could develop. In addition due to the prominence of Massachusetts ports in the shipment of tradable goods Boston became a major centre of wealth, with 5 per cent of Boston's merchants owning 44 per cent of the city's taxable property in 1771. While the ports were of vital significance to the voyagers and merchants of the time it was not until 1807 and 1810 when embargoes were placed upon the trading of goods that merchants were forced to find alternative forms of investment. This factor, along with the war of 1812 and the desire of the Commonwealth to attain independence from Great Britain led to the establishment of the Boston Manufacturing Company in 1813 (Ferguson and Ladd, 1988) which consisted of a textile mill with all the component parts of production being located under one roof. In response to the unprecedented success of this mill, which was located in Waltham, numerous other mills, dependent upon hydro-electric power followed.

The development of other sea ports in the United State, the extraction of mineral ores, the availability of venture capital and a declining supply of cheap labour due to the restricted access of immigrants, all greatly affected the success of the textile industry. Following the Second World War, textile employment in New England fell from 280,000 in 1947 to 170,000 in 1954 and hence to 99,000 by 1964 (Frankel and Howell, 1985). Further, this decline in textile and other traditional manufacturing employment was accentuated during the 1950s and 1960s by capital flight with firms preferring to locate in the South East region of the US where wage costs were lower and the labour was largely non-unionized (ibid.).

The constraining influences upon the traditional industries continued throughout the 1960s, and by the early 1970s unemployment was rising at an unprecedented level. In 1960 the number of people on state welfare was 170,000, yet by mid 1972 this had risen to in excess of 750,000 (Lampe, 1988:61). In addition, between 1969 and 1979 while the level of poverty within the nation as a whole increased by 9.5 per cent, 'official' poverty within Massachusetts increased by 11.6 per cent over the same period (Ross and Trachte, 1990:178). The changing economic environment of Massachusetts has been characterized by three major influences since the Second World War namely, the decline of the traditional and mature industries; the growth of the high technology industry within the region; and the role of military contracts in the development of the state. The main features of these influences are now outlined.

The decline of the mature industries within Massachusetts

'In 1950, the so-called mature industries, which include shoe and leather, apparel, industrial machinery, paper and furniture industries, as well as others, comprized 86 per cent of manufacturing jobs. By 1982, they accounted for 62 per cent of a shrinking manufacturing employment base. The change entailed the loss of 200,000 jobs in the older industries' (Ross and Trachte, 1990:177).

Bennett Harrison of Massachusetts Institute of Technology (MIT) was commissioned by Senator Alan McKinnon in January 1974 to undertake an investigation into the economic problems of the state. In his report he concluded that there were three over-riding factors which were affecting the performance of Massachusetts, namely the collapse in demand for state products, the secular decline in mill-based industries and the geographical location of the region at the end of the national power and transportation networks.

From 1965 to 1972, employment declined within the furniture industry with similar declines within the food processing industry being apparent from 1954. The textile industry had experienced a reduction in employment since 1950, with the apparel industry reporting contractions in growth since 1964. Harrison (1988) points out that in some cases these reductions in employment were due to the introduction of more sophisticated machinery, which led to improvements in worker productivity, yet in most of the cases the decline of the industry was directly related to a growing lack of competitiveness.

The traditional industries, sensitive to cost changes were very susceptible to the relocation of industry during the 1970s and 1980s. Textile, shoe, leather and metal-working firms began to relocate to the South East states of the United States where labour was cheaper and non-unionized. In addition, the New International Division of Labour was leading to an increase in the number of imports, and this challenged the competitiveness of the traditional industries with lower costs of production, and often higher quality leading to the import substitution of such goods as 'shoes and leather, textile machinery, textiles and apparel' (Ross and Trachte, 1990:177).

Bluestone and Harrison (1982) in their examination of deindustrialization within America emphasized that where efficient planning was not undertaken to deal with the maturing industries the costs of economic dislocation would be 'exclusively borne by the work force in that plant, company, or industry, or by the communities where the plants are located' (1982:248). With the re-election of Dukakis as Governor of the state in 1983 it was realized that the problems with the traditional and mature industries had to be resolved. Harrison (1974 as reproduced in Lampe (ed) 1988) also illustrated that the rising number of plant closures stimulated the availability of large and cheap space wherein Massachusetts business could locate. In addition, escalating rates of unemployment produced a reserve of labour which new firms could utilize.

In 1983, a Commission was established on the Future of Mature Industries, comprising 38 members from business, labour, government and academia. The Commission's mandate was to 'develop a comprehensive and cooperative strategy among leaders from business, labor and government for supporting and strengthening the Commonwealth's mature industries' (Dukakis and Kanter, 1988:66). This objective was however, largely bypassed with attention being focused upon the legislation relating to pre-notification of plant closure and widespread lay-offs[3]. (This emphasis upon the latter factor may have been due to the high representation of the Associated Industries of Massachusetts (AIM) and the Massachusetts High Technology Council (MHTC) members on the Commission). Subsequently, the majority of the policies arising from the report of the Commission related directly to the pre-notification legislation. For example, it was agreed that, 'the plight of dislocated workers deserved attention but… that mandatory pre-notification was untenable. [Thus] business might agree to 'voluntary appropriate action,' [or in other words] voluntary pre-notification and voluntary severance benefits. Displaced workers would be guaranteed a package of benefits from their employers or from the state if their employers were unwilling or unable to comply. These benefits included a combination of 90 days notifications and supplemental employment benefits, extension of health care for 90 days beyond termination, and extension of state job training and relocation programs.' (Dukakis and Kanter, 1988:67).

Although the Commission did not formulate policy to stimulate the development of mature industries, it did *attempt* to overcome the social problems associated with plant closures. In addition, through the refinements to the plant closure bill a further step was taken in creating a positive business environment to which businesses would be attracted. Therefore, the attempts by the second Dukakis administration to revive the mature industries of the state merely served the demands of the business community in terms of reducing the legislation concerning their geographical mobility. Yet, the decline of the mill-based industries did enable the location and expansion of high technology and other businesses within large, cheap and accessible buildings. This combined with the reduction of property taxes - such as that of Proposition 2½ - encouraged firms to locate within the region, an example being Digital Corporation which located within an old woollen mill in Maynard (Mullen, Armstrong and Kavanagh, 1986).

While the state aspired to establish policies conducive to the development of the mature industries, the revival of these firms could only be achieved through the re-establishment of a competitive advantage over competing firms (many of which were now located in development countries). Hence, it is proposed that the mature industries should 'encourage specialization, the expansion of business and labor flexibility, and the identification and further development of highly localized, invisible cost advantages' (Doeringer and Terkle, 1990:1304).

Indeed, the Office of Technology Assessment (OTA:1984) concerned with the demise of the traditional industries examined the introduction of computerized manufacturing automation and the effects of this upon employment education and the work place. A number of important observations were noted including the following: 'Programmable automation [PA] is an important and powerful set of tools, but it is not a panacea for problems in manufacturing...in addition to using automation, other steps that need consideration by management include redesigning products for more efficient production, minimizing inventory levels, and improving job design and labor relations' (OTA, 1984:3). In terms of the effect of programmable automation upon the labour force they concluded that, 'A prerequisite of PA-related instruction of all types is a strong foundation of basic skills - particularly reading, science, and math...Development of multiple skills and the *cross-training* of workers to perform a variety of functions on the shop floor are emerging instructional requirements for automated facilities. [Hence] there is an immediate need for retraining and job counseling programs geared to the unique needs of displaced workers' (OTA, 1984:11).

While the OTA recognized the ability of programmable automation to transform the work place and regain some of the competitive advantage which the traditional industries had lost, the introduction of such was not seen as the ultimate goal of the business community. Instead, broader issues were addressed and the realization was explicit that due to the revolutionary nature of this technology an overall restructuring of the work place would be essential in terms of education, training and retraining, if programmable automation was to be successful.

The growth of high technology and the role of military contracts within the state

While the growth of the high-technology industry in Massachusetts was symbiotically related to the allocation of military contracts within the state, the former was not dependent upon the latter. The Centre for Labor Market Studies in 1985 stated that, 'For a variety of reasons, product demand has weakened for producers of computers and electronic components, thus causing employment to decline. At the same time, the data suggest that national defense-related spending has generated numerous employment opportunities in defense-related elements of the high technology sector...' (CLMS as quoted in CLMS, 1991:92). While high technology products penetrate both the consumer and the defense-related markets these are not mutually exclusive with respect to their manifestation within the industry.

Markusen has been instrumental in highlighting the link between high technology and the defence related industry. She outlines that the predominance of high technology firms within Massachusetts, as characterized by Route 128, is a direct reflection of the disproportionate number of military contracts attained within the Massachusetts region, relative to its population. Within Massachusetts the counties of Middlesex and Essex across which Route 128 extends were ranked sixth and twelfth in terms of the total value of defence prime contracts in 1984 relative to the rest of the United States. In addition, Markusen also states that the rise of the high technology defence sector was strongly affected within New England by the 'exit of other lead industries and a deep regional postwar recession' (Markusen, 1991:406).

The concentration of high technology firms within the state of Massachusetts has been attributed to a number of factors. For example, the highly specialized and advanced research institutions of the region such as MIT have been of paramount importance to both the development of modern warfare technology as well as the spin-offs which have emerged from institutional research. In the spring of 1941 the Office of Scientific Research and Development (OSRD) was created to oversee the work of the National Defense Research Committee (NDRC) and the Committee on Medical Research. Although the OSRD did not perform any research it decided on the research to be performed and awarded the contracts for such. By the end of the war the OSRD had spent over $425 million on research and development. Rosegrant and Lampe (1992:93) state that, 'The military's keen interest in maintaining a technological edge, whetted by the experience of OSRD during the war, was sharpened further by the Soviet Union's detonation of its first atomic bomb in 1949, the Korean conflict in the early 1950s, and the successful launch of Sputnik by the Soviets in 1957. As a result, funding for R&D in leading-edge technologies soared and the technical expertise needed to produce better early warning systems, sophisticated new guided missiles, advanced aircraft, and a host of other weapons and defense measures was concentrated in such areas as Cambridge [MIT]'.

During the 1950s, Massachusetts firms were granted over $6 billion in prime contracts from the US Department of Defense (DOD), with this level rising to over $1 billion annually during the 1960s (Rosegrant & Lampe, 1992). The expertise within MIT was manifest in the attainment of such military contracts as the design of the Mark-14 gunsight, the development of an airborne 'inertial guidance' system, and guidance systems for the Titan ICBM, Polaris and Trident missiles. The concentration of military spending in the R&D complex of Massachusetts, led to the development of 'microwaves, electronics, computers, guidance systems, and artificial intelligence' (Rosegrant and Lampe, 1992:100).

While military contracts have been instrumental in stimulating R&D within the high-technology sector, on three occasions since the Second World War, the reduction in military contracts within Massachusetts has led to the high-technology sector examining its vulnerability with regard to military-related contracts from the government. At the end of the Second World War Vannevar Bush, head of the OSRD, aware of the need to apply military research to peacetime efforts acknowledged that, 'Basic research is performed without thought of practical ends... The scientist doing basic research may not be at all interested in the practical applications of his work, yet the further progress of industrial development would eventually stagnate if basic scientific research were long neglected' (Bush, 1945 as quoted in Rosegrant and Lampe, 1992:102). Hence this formed the basis for the foundation of the National Science Foundation (NSF) which stimulated basic research.

A similar period occurred in the late 1960s, when the National Aeronautics and Space Administration (NASA) reduced its spending as the end of the Vietnam War. While real spending on defence fell by 38 per cent between 1968 and 1976 it was the ending of the Vietnam War in conjunction with contractions in space programs and 'unprecedented foreign competition in manufacturing' (Campbell, 1992:7) which led to levels of unemployment similar to those during the 1930s Depression.

During the early 1970s Massachusetts experienced a number of recessions with employment within the manufacturing sector declining by 11.5 per cent in 1974-75 leading to an unemployment rate of 11.9 per cent. Again, at the end of the 1980s - following injections of approximately $150 billion into New England firms during the 1980s by the Carter/Reagan administrations - the decline of the Cold War once again rendered military-related firms vulnerable to changing market conditions. The Federal Reserve Bank of Boston (Campbell, 1992:7) estimates that by 1997 defence cuts could reduce the regional employment level within New England, from the 1991 level by 1.4 per cent or 93,000 jobs, which is approximately double that of the national decline and serves to emphasize the disproportionate concentration of high technology firms, which are dependent upon military contracts within New England and especially Massachusetts.

If these firms are to survive they will have to find new markets through the commercial application of their products and expertise. Although numerous examples of such conversions have been documented such as the movements of the aerospace industry into mass transit, many of these changes have resulted in the closure of the firms concerned. Campbell (1992) in his analysis of Boeing-Vertol[4] states that the problems in conversion as encountered by Boeing-Vertol were typical of other defence contractors. 'Defense firms emphasized performance for extreme conditions, rather than cost, timeliness, or simplicity. Their technologies likewise grew distant from commercial norms. When it came to management, the firms insisted upon a rigid hierarchy, and they lacked the flexibility that commercial businesses demand' (Campbell, 1992:8).

In the past conversions between military and commercial functions have been achieved by the high technology sector within Massachusetts. Although a number of firms have failed to make the transition successfully, over time, the recessions which have been inaugurated through the cuts in defence spending have been succeeded by a revitalization within the state economy. For example, between 1975 and 1980 the manufacturing sector increased its employment levels by 95,000, or 16 per cent of which 80 per cent was directly attributable to high technology firms. Indeed employment in all manufactured durable goods within Massachusetts rose by 91,000 jobs, a rate twice that of the national level (The First National Bank of Boston 1981 as reproduced in Lampe (ed) 1988). In the 1990s, the manufacturing industry within Massachusetts is over-dependent upon the high-technology sector, which accounts for over 40 per cent of manufacturing employment. In September 1991 it was stated that 'the high technology sector of the Massachusetts economy has been perceived since the mid 1970s as a primary engine of economic growth for the state. Composed of both manufacturing and non-manufacturing producers who employ high proportions of scientific, engineering, and technical personnel within their organisations, the high technology sector has been at centre stage for economic development policy in the state for many years' (DET, 1991a:1).

During the 1980s, however, high technology firms have diverged in two distinct directions with the manufacturing sector of the industry continually declining in employment from its peak in 1984. This occurring in a period of sustained economic expansion, leads to the conclusion that prospects for the foreseeable growth in high tech manufacturing is very poor especially with the reduction in military-related spending at the end of the Cold War. Conversely, during the 1980s the high technology service industry has been growing rapidly within the state. The DET (1991a; 1992b) in its economic outlook for the years 1991 to 2005 anticipates that as the economy emerges from the recession of 1990-91 the high technology service industry will be a major force in prompting economic growth.

Economic measures of change in Northern Ireland and Massachusetts

'The industrial structure of an economy can provide some important clues about its potential for growth and its capacity for wealth generation. Industrial structure can also determine an economy's resilience or vulnerability to external shocks' (Murshed et al, 1993:47).

Having outlined the dominant forces at work within the economies of Northern Ireland and Massachusetts, the extent to which the industrial structure within the regions has changed is now examined. Changes within the industrial structure illustrate the inherent strengths and weaknesses within the regional economies and the major sectors of growth over the 1973 to 1994 period. The manufacturing sector is then examined with a relative decline in manufacturing within both regions being found. The relative importance of manufacturing in the regional economies is then examined by manufacturing sector. Herein significant differences between the regions are identified with Northern Ireland displaying a continued dependence upon traditional medium price range products which are subject to intense international competition. In contrast, within Massachusetts the decline of traditional industrial sectors such as apparel and leather goods is counteracted by growth within machinery, electrical and electronic machinery and equipment, and instruments and related products. Having examined the relative proportion of employment within each of the industrial sectors the final part of this section outlines the changing pattern of unemployment within both regions.

Industrial structure

While the industrial structure and the importance of different sectors to regional performance may be measured by output, analysis by employment overcomes differences between industrial sectors in both capital intensities and output to employee ratios. Employment measures also illustrate the importance of different sectors to the local economy and excludes external influences such as remitted profits (Murshed et al, 1993). The industrial structures of the Northern Ireland and Massachusetts economies were therefore examined through employment criteria (Table 2.1).

Within both Northern Ireland and Massachusetts the 1973 to 1994 period witnessed the relative decline of the manufacturing sector along with the growth of the service sector. Yet in Northern Ireland, despite service sector growth, employment within the manufacturing sector still exceeded the service sector in 1994. While the largest sector within Northern Ireland was the Government sector employing 31.8 per cent of total non-agricultural employment, this sector accounted for only 13.6 per cent of non-agricultural employment within Massachusetts in 1994.

Table 2.1
Employment by industrial sector as proportion of total non-agricultural employment, Northern Ireland and Massachusetts 1973, 1985 and 1994

	Northern Ireland	Massachu -setts	Northern Ireland	Massachu -setts	Northern Ireland	Massachu -setts
	1973		*1985*		*1994*	
Mining	2.3	*	2.2	0.0	1.3	0.0
Construction	9.2	4.4	5.9	3.6	4.4	2.9
Manufacturing	35.2	27.2	21.8	23.0	18.8	15.6
Transport & Public Utilities	4.8	5.3	4.0	4.3	4.0	4.4
Wholesale & Retail Trade	12.5	22.2	13.4	23.0	15.5	22.9
Finance, Insurance & Real Estate	5.0	5.7	7.0	6.3	8.1	7.2
Services	7.4	20.2*	13.0	26.6	16.2	33.3
Government	23.6	15.1	32.7	13.1	31.8	13.6
Total Non-Agric. Employment	100	100	100	100	100	100

Note: * Employees in mining were included with employees in services in Massachusetts, until 1983.

Source: Regional Economic Model NIERC; Bureau of Labour Statistics

The most rapid growth in the public sector in Northern Ireland occurred between 1970 and 1980 when employment rose by an average of 4.2 per cent per annum. Although this rate slowed considerably during the 1980s and into the 1990s the public sector still accounts for a higher proportion of total non-agricultural employment than within any other region of the UK. It is the large size of the Government sector which has acted as a counter-cyclical influence in buffering the Northern Ireland economy from recession, as was found in the 1990 to 1992 recession. Yet, the over-dependence upon the public sector also constrains growth during recovery periods.

While Northern Ireland has approximately 16.2 per cent of its employment in the service sector, growth over the 1973 to 1994 period in this sector has been predominantly in consumer services as opposed to producer services. In particular the failure of the Northern Ireland economy to develop financial and business services is cited as the main constraint in the wider restructuring of the region, with the lack of producer services reinforcing the decline in the manufacturing sector (Murshed et al, 1993).

In contrast to Northern Ireland the service sector accounts for a third of all non-agricultural employment within Massachusetts. It is the service sector within Massachusetts which has grown most rapidly from 1973 to 1994 with the only other growth sector being the financial, insurance and real estate sector. Tradable services in particular have growth rapidly and this was closely aligned to the growth in the high-technology sector. Within the Massachusetts economy it was the wealth creating sectors which were expanding, such as in producer services and high-technology as opposed to the government sector.

Manufacturing

The relative change in manufacturing employment relative to total non-agricultural employment for Northern Ireland and Massachusetts, over the 1973 to 1994 period is illustrated in Figure 2.1. Throughout the 1960s total manufacturing output remained stable within Northern Ireland, although the proportion of civilian employment in manufacturing declined as capital was substituted for labour. From the early 1970s and the oil crisis of 1973, the relative share of manufacturing to Gross Domestic Product (GDP) and employment declined within Northern Ireland. In 1973 manufacturing accounted for 35.2 per cent of total non-agricultural employment (and 33.2 per cent of total employment), yet by 1994 this had declined to 18.2 per cent. Similarly, while manufacturing contributed 27.6 per cent of total GDP in 1973, this had declined to 20.6 per cent by 1994.

In Massachusetts 27.2 per cent of non-agricultural employment was found within manufacturing in 1973, yet like Northern Ireland manufacturing's share of employment declined throughout the 1970s and 1980s. By 1994, the manufacturing sector accounted for only 15.6 per cent of total non-agricultural employment. Within both regions therefore the manufacturing sector has declined in importance relative to the other industrial sectors.

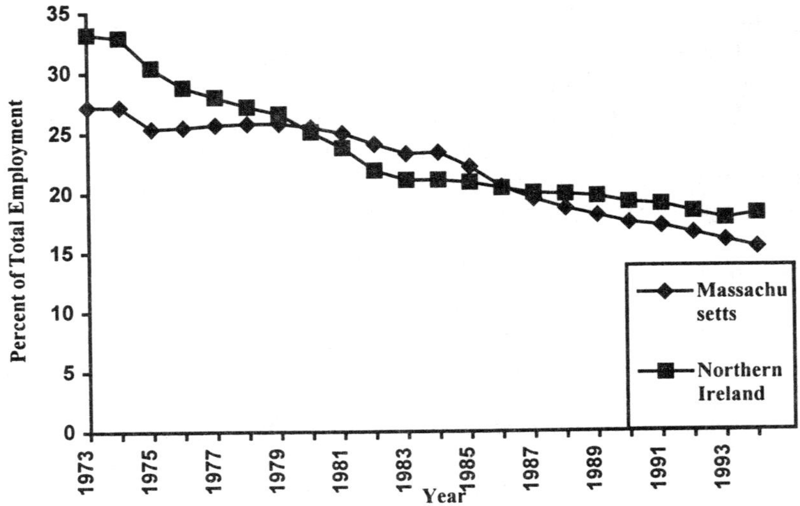

Figure 2.1 Employees in manufacturing as a percentage of total non-agricultural employment 1973 to 1994

Manufacturing sectors

Despite the declining importance of manufacturing relative to other industrial sectors since the 1970s, the performance of the manufacturing sector still remains central to regional development[5]. The relative importance of the different manufacturing sectors within both the Northern Ireland and Massachusetts economies was measured in terms of employment.

Within Northern Ireland the relative importance of each of the manufacturing sectors based upon their share of total manufacturing employment has remained relatively unchanged between 1973 and 1994 (Table 2.2). In both 1973 and 1994 the textiles sector and the food, drink and tobacco sector accounted for the largest share of manufacturing employment. While employment levels in the textile sector in 1994 were only 48.2 per cent of their 1973 level (Table 2.3), this sector still accounted for 24.1 per cent of total manufacturing employment in 1994. Within the food, drink and tobacco sector although the share of total manufacturing increased from 15.6 per cent in 1973 to 19.6 per cent in 1994, employment within this sector was only 77.3 per cent of the 1973 base level. Therefore while the absolute number of employees within food, drink and tobacco declined over the 1973 to 1994 period, relative to the other manufacturing sectors, food, drink and tobacco increased in importance. The only sector to experience employment growth between 1973 and 1994 was the wood products sector (Table 2.3), which rose from the tenth to the seventh largest share of manufacturing employment in 1994. This continued dependence upon traditional industries with mid-price range products, and the failure to develop science based industries has made the Northern Ireland economy vulnerable to low cost competition. The over-dependence within Northern Ireland on industries which are contracting within Europe would therefore lead to questions concerning the viability of the structure of the manufacturing sector.

In the mid-1990s manufacturing in Massachusetts was responsible for the employment of approximately 17.5 per cent of the states labour force. Within Massachusetts those manufacturing sectors which accounted for the largest proportion of manufacturing employment in 1973, were still the dominant employers in 1994, most notably the machinery and electrical and electronic machinery and equipment sectors which include 'computers, electronic and electrical equipment such as telephone apparatus, semi-conductors and light bulbs; precision instruments for cameras, hospitals and guided missiles' (DET, 1991:10) (Table 2.4). While the majority of the manufacturing sectors maintained their relative ranking there were two exceptions to this pattern. Within the durable goods sector instruments and related products increased their relative share of manufacturing employment from 1973 to 1994, and rose from sixth to equal second place in relative importance. In contrast, the non-durable, apparel and other finished products sector reduced the share of manufacturing employment over the period, falling from the fourth largest manufacturing employer to the 12th rank. These results demonstrate that while machinery, electrical and electronic machinery and equipment dominated manufacturing employment in the 1970s and 1980s, these sectors still remain as the main employers in the 1990s. The traditional sectors of apparel and other finished products and leather and leather products have declined significantly since 1973 while instruments and related products have improved their relative share of manufacturing employment within Massachusetts.

Table 2.2
The contribution of each manufacturing sector to total manufacturing employment in Northern Ireland, 1973 and 1994

	Percentage of Manufacturing Employment	Rank	Percentage of Manufacturing Employment	Rank
	1973		*1994*	
Manufacturing Sector				
Food, Drink & Tobacco	15.6	2	19.6	2
Textiles	30.8	1	24.1	1
Wood Products	3.2	10	5.6	7
Pulp, Paper & Printing	3.9	8	6.4	6
Coke, Oil Refining & Nuclear Fuel	0.1	13	0	13
Chemicals & Man-Made Fibres	6.9	6	4.2	9=
Rubber & Plastic Products	4.5	7	4.6	8
Other Non-Metallic Products	3.7	9	4.2	9=
Metals	2.8	11	2.2	11
Machinery & Equip.	8.2	5	7.9	5
Electrical & Optical Equip.	8.7	4	8.8	4
Transport Equip.	10.6	3	11.1	3
Other Manufacturing	0.7	12	1.0	12
Total	100		100	

Source: Regional Economic Model

Table 2.3
Employment change in industrial sectors as a proportion of total manufacturing employment 1973 to 1994, Northern Ireland

	1973	*1975*	*1980*	*1985*	*1990*	*1994*
Manufacturing Sector						
Food, Drink & Tobacco	100	96.8	90.6	82.2	74.9	77.3
Textiles	100	86.8	70.3	53.0	50.6	48.2
Wood Products	100	91.9	95.6	97.0	101.8	107.2
Pulp, Paper & Printing	100	95.2	93.3	85.9	93.7	99.5
Coke, Oil Refining & Nuclear Fuel	100	100	100	0.0	0.0	0.0
Chemicals & Man-Made Fibres	100	101.1	71.4	32.5	34.1	37.5
Rubber & Plastic Products	100	87.0	89.0	45.6	59.1	62.8
Other Non-Metallic Products	100	93.9	91.5	76.3	70.6	69.6
Metals	100	74.7	74.0	55.7	51.9	49.6
Machinery & Equipment	100	98.4	81.4	60.8	62.1	59.7
Electrical & Optical Equip.	100	105.0	74.1	55.3	64.9	62.1
Transport Equip.	100	102.5	91.6	83.2	73.0	64.1
Other Manufacturing	100	98.8	102.5	90.8	102.2	88.8
Manufacturing	100	94.1	80.8	63.8	62.7	61.6

Source: Regional Economic Model

Having established the relative importance of the different manufacturing sectors in 1973 and 1994 the absolute changes within the different Massachusetts sectors were examined (Table 2.5). Absolute changes in employment within each of the manufacturing sectors using 1973 as the base year demonstrates that relative and absolute changes were not always complementary.

Within Massachusetts the instruments and related products sector demonstrated the largest percentage increase in employment from 1973. This was closely followed by the printing and publishing sector whose employment was 156 per cent of the base year level. Additionally, the chemicals and allied products sector also increased its employment to 125 per cent of the 1973 level. Although the machinery sector improved its relative position over the period and was the largest manufacturing employer in 1994, its employment growth had not been as large as the instruments and related products, the printing and publishing, and the chemicals sectors. The relative decline of the apparel sector was reflected in the 1994 employment level representing only 51 per cent of the 1973 level. Further, while the electrical and electronic machinery and equipment sector was the second most important sector in 1994 it had experienced variable employment growth rates over the 1973 to 1994 period, with the 1994 level representing a 5 per cent fall on the 1973 base.

Table 2.4
The contribution of each manufacturing sector to total manufacturing employment in Massachusetts, 1973 and 1994

	Percentage of Manufacturing Employment	Rank	Percentage of Manufacturing Employment	Rank
	1973		*1994*	
Manufacturing Sector				
Lumber & Furniture	2.5	16	1.8	15 =
Stone, Clay & Glass	2.2	17	1.8	15 =
Primary Metal Products	2.6	15	2.1	14
Fabricated Metal Products	8.8	3	8.0	5
Machinery (except Elect.)	11.5	2	14.1	1
Electrical, Electronic Machinery, Equipment etc	13.9	1	13.1	2 =
Transportation Equipment.	5.2	8 =	4.3	9
Instruments & Related Products	6.8	6	13.1	2 =
Misc. Manufacturing	4.4	12	3.8	11
Food & Kindred Products	5.0	10	4.6	7
Textile Mill Products	4.9	11	3.4	13
Apparel & Other Finished Products	7.2	4	3.7	12
Paper & Allied Products	5.2	8 =	4.5	8
Printing & Publishing	6.9	5	10.7	4
Chemicals & Allied Products	3.2	14	4.0	10
Rubber & Misc. Plastic Products	5.5	7	5.5	6
Leather & Leather Products	4.2	13	1.2	17
Total (Per Cent)	100		100	

Source: Massachusetts Department of Employment and Training

Table 2.5
Employment change in industrial sectors as a proportion of total manufacturing employment 1973 to 1994, Massachusetts

	1973	*1975*	*1980*	*1985*	*1990*	*1994*
Manufacturing Sector						
Lumber & Furniture	100	88	78	81	69	73
Stone, Clay & Glass	100	84	87	79	75	80
Primary Metal Products	100	105	101	95	85	82
Fabricated Metal Products	100	96	89	79	89	90
Machinery (Except Electrical)	100	111	141	159	142	122
Elect, Electronic Machinery, Equipment Etc.	100	103	118	123	101	95
Transportation Equipment	100	99	105	122	102	83
Instruments & Related Products	100	112	128	132	197	193
Misc. Manufacturing Industries	100	94	89	70	76	87
Food & Kindred Products	100	100	82	72	78	93
Textile Mill Products	100	89	77	59	57	70
Apparel & Other Finished Products	100	99	81	68	51	51
Paper & Allied Products	100	92	80	76	83	86
Printing & Publishing	100	102	95	120	146	156
Chemicals And Allied Products	100	107	85	85	106	125
Rubber & Misc. Plastic Products	100	88	83	83	84	101
Leather & Leather Products	100	94	75	46	32	29

Source: Massachusetts Department of Employment and Training

Unemployment

Irrespective of the business cycle, unemployment within Northern Ireland has been consistently higher than the Massachusetts rate (Figure 2.2). Over the 1973 to 1994 period average unemployment in Massachusetts has been 6.4 per cent as compared to 12 per cent within Northern Ireland. While comparability is found between the Massachusetts and US unemployment rate over the 1973 to 1994 period, the Northern Ireland rate of 12 per cent is considerably higher than the UK average rate of 7.2 per cent. Although changes in unemployment during the 1990s have been evenly distributed throughout the UK, Northern Ireland continues to have the highest unemployment rate. Although the gap between Northern Ireland and the UK narrowed quite rapidly during the recession years of the early 1990s, from a peak of 7.4 per cent in 1990 to 3.5 per cent by 1993, the difference between the average UK rate and the Northern Ireland rate since then has remained relatively stable at 3.5 per cent. The majority of the gain in Northern Ireland's relative unemployment position occurred in 1991 due to the large falls in GDP in the Southern and Midlands regions of the UK. This therefore reduced the overall disparity in unemployment rates for Northern Ireland.

The performance of the Massachusetts economy has changed significantly from the 1970s to the 1990s. In 1975 the state had an unemployment rate of 11 per cent, yet, by 1987 this had declined to 3.2 per cent. Further, in 1975 the state had a trade deficit of $600 million yet by 1985 it was in surplus by $950 million. While the unemployment rate in Massachusetts has been consistently lower than the US average since 1978, in 1990 it rose above the national average. This increase in the Massachusetts unemployment rate is attributed to changes in the industrial base at the end of the 1980s. In 1990, 'the industries that generated the bulk of the jobs - construction, services, retail trade, finance, insurance and real estate - declined' (DET 1991:4). By 1994 however, unemployment in Massachusetts was equivalent to the US average of 6 per cent.

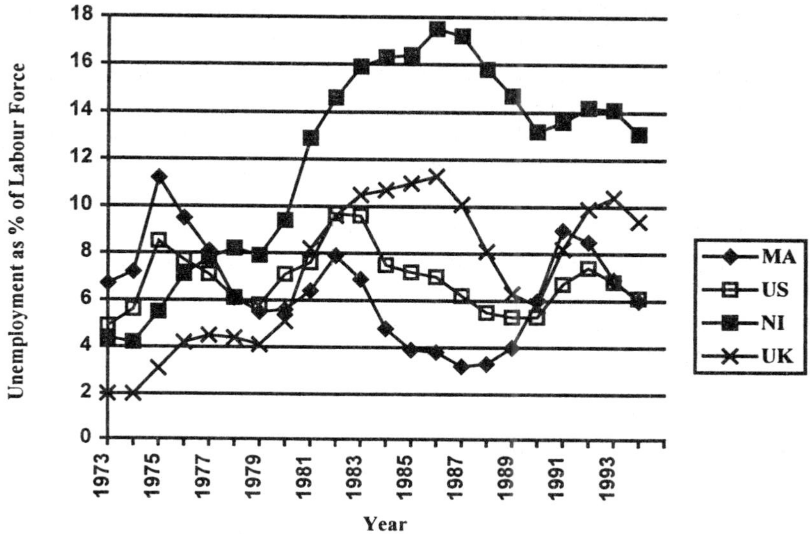

Figure 2.2 Unemployment rates as a percentage of the civilian labour force, 1973 to 1994

Source: Regional Forecasting Model, NIERC; Bureau of Labour Statistics, Massachusetts

The small firm sector in Northern Ireland

From the end of the Second World War until the early 1970s, economic policy within Northern Ireland was moulded by policy initiatives at national level and was heavily dependent upon branch plants. In the early 1970s world economic conditions were deteriorating, and the oil crisis of 1973 culminated in the contraction of industries and production throughout the world. As a result of the inherent character of branch plants - that is, being concentrated predominantly in the low order functions such as production, while the higher order functions, namely Research and Development, and higher order managerial work are located in company headquarters - it is commonly found that during a period of recession, it is the branch plants which are the first to close. Within Northern Ireland the closure of a number of branch plants between 1973 and 1986 resulted in the loss of 50,000 jobs. Evidence suggests that it was those large firms which moved into the province under the auspices of the ID legislation which were most severely affected, falling in absolute numbers of firms from 134 to 65 over the period 1973 to 1986, while employment declined by approximately 65 per cent, to 10,000 employees (NIERC, 1989:34).

Industrial policy was faced with two options namely, the re-attraction of inward investors, or the nurturing of the indigenous sector. The dilemma in the choice of the latter, was that it was a long-term strategy and would take many years of fostering before employment gains would be evident. Conversely, the successful attraction of just a few large branch plants would temporarily solve the crisis of unemployment in Northern Ireland. While the attraction of inward investors as an economic strategy was not abandoned, it was the nurturing of the indigenous sector which was more fully embraced as a key to economic revival, albeit in the long-term. Indeed, the chosen path was not selected as the most suitable or advantageous, but rather, it was adopted in an era when the economy of Northern Ireland was proving unable to attract inward investment due to rising social and political unrest.

The Republic of Ireland resembled Northern Ireland in that it wanted to concentrate upon the indigenous sector and regarded a policy of dependence upon inward investment as foolish (Telesis, 1982, as highlighted in Hitchens and Birnie, 1992:54). Yet, the Republic Of Ireland often faced insurmountable problems in breaching the risk-aversive attitudes which were indoctrinated within the cultural norms of the people, and thus the widescale provision of financial assistance, both directly and indirectly, to inward investors prevailed throughout the 1970s and 1980s. Consequently, by 1986 the Republic Of Ireland had attracted 43,076 new jobs due to companies moving into the country since 1973. In contrast Northern Ireland only attracted 6,888 jobs through this method over the same time period.

Within Northern Ireland job promotion was central to policy initiatives in the 1970s and 1980s as evident in the recommendations of the 1970-75 Development Programme and the Wilson Plan. While the number of new VAT registrations has risen faster in Northern Ireland than in Great Britain since 1979 the increase in the number of small firms has been most marked in the manufacturing sector. Between 1979 and 1993 the number of firms in the production sector rose by 65 per cent in Northern Ireland as compared to 18 per cent in Great Britain (UK VAT Statistics, Department of Employment, Various Years). There has been a disproportionate increase in the number of small businesses in Northern Ireland as compared to other regions of the United Kingdom since the 1970s. Northern Ireland manufacturing is now largely dependent upon the small firm sector, which has provided strength through diversity (Gudgin et al., 1995) and reduced the vulnerability of the economy to the location/relocation of branch plants.

The Local Enterprise and Development Unit (LEDU) which has responsibility for small firm development within Northern Ireland has been instrumental in stimulating the small firm sector. From the formation of LEDU it quickly created new jobs, although its major impact was not felt until 1981 when job promotions increased rapidly[6]. Gudgin and Hitchens (1990:137) illustrate that in 1973 only 12 per cent of manufacturing employment was found in small firms - that is, those firms employing under 50 employees - but by 1986, this level had increased to 24 per cent, largely due to the work of LEDU. Although a number of the new firms had closed between 1973 and 1986 there was a net increase of 9,800 jobs.

It is perhaps with reference to the record of LEDU in the past, as it has concentrated on generating job promotions often in the short-term, as projects funded have been deemed unsuitable resulting in their eventual closure, that in the 1990s emphasis has now moved away from job promotions, to job expansions within firms which have the potential to grow. This will necessitate a closer inspection and analysis of factors beyond the viability of an idea, to be turned into a business, to the consideration of factors such as market share, price, quality, managerial expertise, competitiveness etc., because only in light of these factors, can an accurate decision be made as to the effectiveness of providing financial and other assistance.

As demonstrated in Figure 2.3, even since 1980, there has been a substantial increase in the numbers of self-employed, however, this has not been equally distributed across the three sectors of agriculture, production and services. Most notably, self-employment within the agricultural sector has declined, while that for production and services has increased. There are numerous factors to account for these changes, yet it is plausible that the introduction of *quotas* through the Common Agricultural Policy has led to the closure of many private agricultural enterprises. Further, the financial and consultancy sectors of the Northern Ireland economy have grown rapidly, which may be a factor in the growth of self-employment within services.

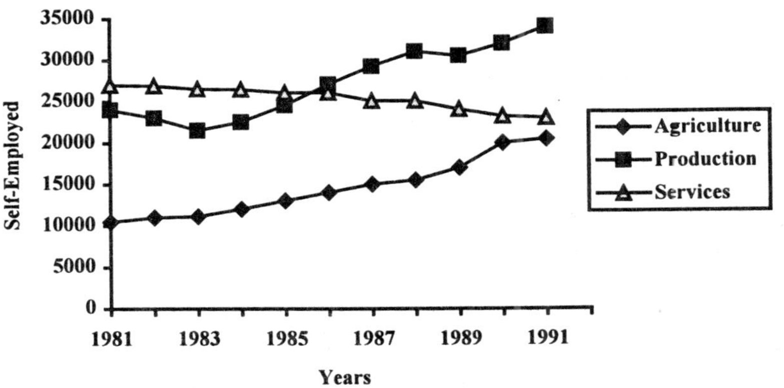

Figure 2.3 Self-employed in Northern Ireland 1981-1991

Finally, the increase in the number of self-employed in the manufacturing sector may be due to LEDU's efforts being directed towards this sector with the aim of building up a strong manufacturing base to stimulate further development within the economy. Additionally, there is the possibility that with the restructuring of the economic base of the national, as well as world wide economies that subcontracting has, and will rise in significance, thus promoting the formation of new small manufacturing firms within peripheral regions such as Northern Ireland.

The small firm sector in Massachusetts

In 1987 Ronald Reagan stated: 'I urge all Americans who own or work in a small business to continue their resourcefulness and successes, for these efforts contribute so much to the entrepreneurial spirit which made this nation great. It is with justifiable pride that the American small business man can point to himself as the backbone of our nation.' (as quoted in Kuratko, 1988:128).

As within the United Kingdom, small businesses have been increasingly recognized as instrumental in economic development within the United States. The Small Business Administration, which holds the legal oversight of all US small businesses, was granted $16.2 billion in 1990/91 from federal coffers, a figure demonstrating the enormity of both the Administration and the importance of a healthy small business sector to national prosperity.

The political emphasis placed upon small firms within the United States escalated following the research by Birch (1979). With regard to the economy of New England Birch's study concluded that 'although small businesses have a far greater failure rate than large ones, the contractions of large companies since 1969 represented a greater share of lost jobs than small business failures. About 40 per cent of replacement jobs come from independent businesses with fewer than 20 employees...and 60 per cent were generated by all businesses with fifty or fewer employees...[thus] it is the willingness of small businesses to form and expand in New England that is keeping the New England economy as strong as it is' (Birch 1983 as reproduced in Lampe (ed) 1988). Birch's investigation of New England businesses concluded that the proportion of small businesses - that is, with 0-99 employees - was similar to the national level of 98 per cent, which employed 48 per cent of the work force.

In light of Birch's findings it was not surprising to find attitudes amongst state and federal bodies in favour of the stimulation and subsequent creation of small businesses. Indeed, in the SBA's 1991 financial report, they claimed that, 'Small businesses are the key to America's future. Their success will depend on a better educated workforce and to a large extent on their recognition and pursuit of new opportunities, such as those in the global marketplace' (SBA, 1991:46).

There was a clear realization within official bodies that if the small firm sector was to prosper, then it would have to be fully equipped with a vibrant, fully educated and skilled workforce, which would maximize the potential of the small firms. During the era of mass production the bureaucratic model of the organisation facilitated the mass employment of semi-skilled workers. Today however, as flexible technologies are commonplace within the manufacturing firm, there is an escalating need for highly skilled workers within smaller sized firms to meet the demands of the new economy. Yet these employees must not only be skilled to carry out their jobs efficiently, but they must also have the education and progressive up-grading of skills to allow the quick and problem free adaptation to new technological demands. Chandler (1977:333 as quoted in Starr 1988:42) stated that, 'markets…and technology (have) had a far greater influence in determining size and concentration in American industry than did the quality of entrepreneurship, the availability of capital or public policy.' It would be anticipated that where economies could be obtained within the firm through the substitution of factors of production, then economies of scale would no longer be the domain of the large firm, with new technologies providing the impetus for economies of scope and the consequent economic justification of the small firm.

With regard to the identified need to advance the skills of employees, the SBA has taken limited action to target employee skills. For example, the SBA states that 'few small businesses are in a position to cover the full cost of investing in basic skills training for their workers. They have limited resources to pay for the management time to plan and set-up training on-site, equally difficult is paying workers to take time away from their duties to participate in training' (SBA, 1991:46). Yet, the SBA has done little to counteract this inadequacy.

Business failures within Massachusetts in the first half of 1992 increased to 1,586 from 1,309 in the same period of 1991, representing a 21 per cent increase. Further, in June 1992, Massachusetts business failures increased by 36 per cent on the previous year. Therefore, the total liabilities for the 1991/92 financial year amounted to $3.6 billion, an increase in liabilities of 32 per cent from the previous year (Dun and Bradstreet, 6/8/92).

Undoubtedly, small companies which are characterized by a small range of products, specialized markets, low assets and a small market share, are the most susceptible to economic shifts whether through changing consumer spending patterns, or tastes and preferences. Additionally, as the small business community frequently acts as a supplier to the larger multinationals, fluctuations within world markets may lead to the closure of large subsidiary plants thus blocking the market for many of their goods. Yet, the proportion of small firms[7] has remained relatively constant within the Massachusetts economy since the 1970s (Table 2.6). Further, the change in the proportion of total employment within small firms relative to all establishments has closely mirrored the changing proportions of small establishments. What is significant from the data is the proportion of small manufacturing firms relative to all small firms between 1978 and 1995. As aforementioned there has been a large growth in producer services since the 1970s in Massachusetts. As the majority of these firms are small, it follows that in the absence of reciprocal increases in the number of small manufacturing firms, this will lead to a proportionate decline in the importance of small manufacturing firms.

The small firm sector within Northern Ireland and Massachusetts has followed different paths since the 1970s. Within Northern Ireland the small firm sector has grown significantly since the 1970s with this being closely associated with active government policy directed at the nurturing of indigenous industry. While within Massachusetts there has been a strong symbiotic relationship between the large and small firm sectors, within Northern Ireland government policy has tended to focus on large and small firms separately. It is widely recognized that the strength of the high technology sector within Massachusetts is a reflection of the agglomerative economies which are present along Route 128. This offers competitive strengths to small and large firms alike in terms of operational efficiency. In contrast, within Northern Ireland separate inward investment and indigenous policies have failed to exploit the benefits for regional economic development of industrial clusters.

Table 2.6
Proportion of establishments and employment in small firms within Massachusetts, 1978 to 1995

	Small firms as percentage of all establishments	Small manufacturing firms as % of all manufacturing firms	Employment in small manufacturing firms as % of all employment in small firms
1978	93.8	79.1	7.4
1979	93.7	78.6	7.0
1980	92.5	76.8	7.4
1981	92.5	77.8	7.5
1982	92.5	78.6	7.5
1983	92.7	78.6	7.2
1984	92.5	78.0	7.1
1985	92.4	78.8	7.0
1986	92.6	78.9	6.5
1987	93.4	80.3	6.0
1988	93.7	80.5	5.7
1989	97.0	78.0	5.1
1990	94.2	81.5	5.3
1991	94.6	82.7	5.3
1992	94.3	81.6	5.1
1993	94.1	81.6	5.2
1994	94.1	*	*
1995	94.1	81.4	4.9

Notes: * denotes statistics unavailable

Source: ES-202 Employment, Contributions and Wages Report, Department of Employment and Training, Massachusetts

Northern Ireland is distinct from Massachusetts, in that while Massachusetts resembles a core region in terms of the supply of labour, technology, research expertise and business services etc. Northern Ireland is characterized by peripherality and therefore must develop its indigenous sector to avoid the marginalising tendencies of branch plant closures during times of recession. Thus it is for Northern Ireland in particular that small firms are central in the development of a seed-bed of innovation and growth. It is for this reason that the matched plant analysis seeks to identify strengths and weaknesses within the indigenous small firms in Northern Ireland against a core region such as Massachusetts.

Conclusions

This chapter has presented a truncated explanation of the main characteristics of economic development in Northern Ireland and Massachusetts from the early 1970s to the mid 1990s.

Massachusetts has experienced the wrath of recession as manifest through unemployment, rising costs of living, and crippling tax rates. The state however, reacted to changing economic conditions over the period in a way which has facilitated its revitalization following the recessions of 1974-75; 1981-82; and 1990-91. In 1975, following the oil crisis, and the reduction in military appropriations at the end of the Vietnam War, Massachusetts was characterized by abandoned textile and shoe factories, and other such industries, which were faced with the problems of the New International Division of Labour (NIDL), through the availability of cheap labour and new technologies within developing countries. Rates of unemployment reached a high of 12.3 per cent in 1975 (DET 1991), although some areas with a high industrial concentration experienced even higher levels, such as Fall River, Worcester and Lowell. By 1985 however the Massachusetts economy reached an unemployment level of 3.8 per cent, a point which economists described as full-employment. The state maintained a level of full employment through to the fourth quarter of 1988, largely as a result of 'continued job growth and a slowing in the expansion rate of the state labour force' (DET, 1990:3). To move to this prosperous position after the depression of 1975, it involved a 30 per cent expansion in the employment base, creating 718,200 new jobs, yet this brought with it a fundamental change in the structure of the industrial base. These changes included, (DET, 1990:4):

1. Emergence of new high technology industries producing computers and related equipment, sophisticated instruments, chemicals and advanced machinery;
2. Strong gains in the construction industry due to rapid increases in the number of new households and also a number of state infrastructural programmes;
3. Continued growth in retail trade due to a greater than proportional increase in disposable personal incomes in Massachusetts;

4. Expansion of insurance, finance and real estate industries due to factors such as government deregulation of banking, increasing consumer credit and the construction industry expansion;
5. Dramatic growth in the service sector including services to businesses i.e. accounting, data processing, educational and health services;
6. Decline in manufacture of durable and non-durable goods due to the Federal Reserve Bank's strong dollar policies, changing consumer tastes and international competition.

Within Northern Ireland the overall economic picture has remained relatively static throughout the past 25 years with the province maintaining its economically disadvantaged status, both within the United Kingdom and the European Union. The increasing peripheralisation of Northern Ireland pervades all sectors of life including the political (as manifest through the political unrest since the late 1960s); social (as evident in high fertility rates and high levels of out-migration); economic (Northern Ireland having the highest unemployment rate within the UK throughout the 1970s and 1980s); and geographic (being physically peripheralized from the core regions of the European Union. With the decline of the traditional industries within Northern Ireland, most notably in food, drink and tobacco and textiles and the reconcentration of inward investors from the region during the 1980s, a change in policy emphasis away from a dependence upon inward investors, towards the nurturing of indigenous enterprises occurred. The Northern Ireland economy has experienced a growing public sector, a declining manufacturing sector, persistently high levels of unemployment and extraordinarily high fiscal receipts of transfer payments from Westminster. While Government intervention has curtailed the unemployment rate and acted as a buffer against recessionary pressures, it has also prevented market forces from highlighting the uncompetitiveness and inefficiencies within the industrial structure. Hence 'in the absence of any need to change, people benefiting from the system have been passive, their silence a vote for the status quo. The result has been a system that constrained both innovation and growth' (Northern Ireland Growth Challenge, 1995:10).

This chapter has demonstrated that economic development in Northern Ireland and Massachusetts economies since the early 1970s has been significantly different. The present industrial structure is a reflection of the economic trajectories which have occurred in the past. Therefore while this research is conscious of the factors at work within the local economies of Northern Ireland and Massachusetts, in terms of the factors which have moulded the present industrial structures, attention is focused upon the inherent characteristics of the small indigenous sector. The development and nurturing of the small firm sector has been identified within Northern Ireland as vital to the economic vitality of the region. Yet, if such a policy is to be successful it is essential that the comparative performance of such firms is examined. For example, if the firms within a region are operating inefficiently - as measured by factors such as productivity and the quality and price of products - it is only through comparisons with similar firms in other regions that the source of competitive strengths and weaknesses can be identified.

The traditional industrial sectors within Massachusetts have acknowledged that the introduction of new technology is not enough to aid their revitalisation. Instead they recognise the centrality of other aspects within their operation, as enveloped within the philosophy of flexibility, in regaining a competitive advantage (OTA, 1984:3). The matched plant analysis within this book therefore examines firms which are structurally similar within an industrial sector which is progressive and fundamental to the local economy due to the high-value added aspect of much of its production, namely engineering; and a traditional industrial sector, food processing.

Notes

1 The Capital Grant of 1954 and the Investment Grant of 1566 ended in 1971 being replaced by the Standard Capital Grant in 1972.

2 Although this document is commonly known as the 'Pathfinder' report, the full title of the document was "Building a Stronger Economy, The Pathfinder Process".

3 Since 1979 a bill of pre-notification had been filed within the legislative office demanding advance notification by plants intending to close or lay-off large numbers of their work forces. The bill had faced widespread opposition, most notably from the Associated Industries of Massachusetts (AIM) and the Massachusetts High Technology Council (MHTC) who together had prevented it proceeding beyond the committee phase. These opposents wished to at least alter the bill, by removing conditions relating to the reinvestment by closing firms in the local communities, the reduction of the legislation relating to severance pay, and the reduction in the time period of pre-notification (Ross and Trachte 1990).

4 Boeing-Vertol was an aerospace company which failed in its conversion from military contracts to the manufacture of light rail cars for Boston.

5 See Dertouzos, Lester and Solow (1989) for the rationale associated with continued analysis of the manufacturing sector despite declining relative output and employment.

6 The extent to which these job promotions were associated with fragmentation strategies by larger firms and the widespread closure of branch plants during this period is unknown.

7 Small firms being defined as those with fewer than 50 employees.

3 Matched plant methodology

There is no merit in equality, unless it be equality with the best, *John Lancaster Spalding*

We are half ruined by conformity; but we should be wholly ruined without it, *Charles D Warner*

Introduction

The matched plant methodology as an analytical tool has been used widely in the measurement of relative productivity performance between different countries. Since the early 1900s businesses have gauged their relative competitiveness and performance against similar firms in other countries. Attempts have been made to identify *best-practice* and emulate the characteristics of this so that equality may be attained. The matched plant methodology emanates from the desire to improve relative competitiveness and through the identification and conformance of businesses to those aspects which have provided firms with a competitive advantage, close the competitiveness gap.

The matched plant methodology has been applied to both the manufacturing and service sectors at regional, national and international levels. Whilst acknowledging the inherent strengths and weaknesses of this methodology, it represents a valuable tool in identifying comparative performance. To date the matched plant methodology has been used almost exclusively to gauge the relative productivity performance of firms within different countries and identify the reasons for this. The use of the methodology within this study departs from the traditional approach and examines the extent to which corporate flexibility is found within the two regions of Northern Ireland and Massachusetts. By controlling for a number of independent variables such as firm size, employment levels and manufacturing orientation of the matched plants an understanding of the differences in internal functioning of structurally and strategically similar firms is determined. In addition, by measuring the comparative growth levels of the firms it is possible to identify whether or not corporate flexibility is positively associated with growth.

Comparative research within Northern Ireland based upon matched plant analyses have been few, limited mainly to that by Hitchens, Wagner, and Birnie (1990) and Roper and Hofmann (1993). These studies have however focused upon productivity and performance differences between European firms as opposed to a European-American analysis. The comparative research presented here between Northern Ireland and Massachusetts overcomes this omission in the research literature and contrasts a region which has been heralded as displaying the characteristics of flexible specialization, Massachusetts, with a peripheral and economically disadvantaged region, Northern Ireland. The state of Massachusetts is characterized by a rejuvenated economic base in the form of high technology agglomerations and the creation of a vibrant business services economy. In contrast, Northern Ireland displays the characteristics of peripheral entrenchment, being overly dependent upon traditional industrial sectors which are in terminal decline throughout Europe and an exceptionally large public service sector. By examining the characteristics of small manufacturing firms within Massachusetts the validity of the flexibility thesis will be tested. Further, by measuring the growth performance of Northern Ireland and Massachusetts firms it will be possible to determine the effects of corporate flexibility on performance. As the matched plant analysis examines specific business functions as opposed to an holistic measurement of performance it is possible to detect inherent strengths and weaknesses within the firms and therefore extract tangible lessons for the Northern Ireland matched plants.

The remainder of this chapter examines the use of the matched plant methodology during the 20th Century and the advantages and criticisms of the approach. The characteristics of the Northern Ireland and Massachusetts matched plants are also outlined in terms of the selection criteria and the composition of the sample. The component parts of the questionnaire are then presented in terms of the nature of the questions within each section and the use of the data in the examination of corporate flexibility.

The evolution of matched plant methods

There has been a growing awareness throughout the 20th Century of the lessons which can be learnt through comparative analysis of structurally and strategically similar firms. As early as 1902, a group of British industrialists from the Lancashire cotton industry visited the United States to compare the organisation of production and performance between the two countries. Since then a number of other matched plant investigations have been performed to ascertain the relative performance of British firms against their American and European counterparts.

Although matched plant analysis during the first half of the 20th Century was predominantly Anglo-American in focus, recent analyses have been mainly European. This internal focus emanates from the formation of the European Union and the creation of a Single European Market on 1st January, 1993. The Cecchini Report (1988) proposes that the relatively poorer performance of European firms as compared to US and Japanese firms will be reversed through the formation of the Single European Market (SEM) and the provision of economies of scope and scale will result in productivity and efficiency gains. Comparative analysis has therefore become increasingly insular with matched plant studies identifying differentials in productivity between core and peripheral regions of the EU.

Despite the growing emphasis on European matched plant analyses, there have been a number of important Anglo-American studies. The most significant was the Anglo-American Council on Productivity (ACCP), which provided the first comprehensive comparative study between the two countries. The ACCP consisted of 66 teams which examined the industrial efficiency within American and British industry and through the analysis of factors both within the firms and the local environments, concluded that 'British industrial productivity must be raised, and can be raised, comparatively quickly' (Hutton, 1953:8).

Further, Rostas (1948), (as outlined in Hitchens et al., 1990) examined 31 industrial sectors within Great Britain and the United States between 1935 and 1937 and although he discovered that productivity was higher within the United States firms for each industrial sector, the differential varied greatly with productivity within the United States food, drink and tobacco industry representing only 156 per cent that of the Great Britain level, whilst the United States engineering and metals sector exceeded the Great Britain productivity level by 250 per cent.

In 1950, Paige and Bombach (1958, as outlined in Hitchens et al., 1990) also found that the industries within the United States outperformed their Great Britain counterparts in each of the 44 industrial sectors which they investigated. Yet, again the productivity gap varied between industries ranging from 111 per cent of the British level in Shipbuilding to 600 per cent in fuel refining (Hitchens, Wagner and Birnie, 1990). A recent international comparison of productivity between Great Britain and the United States was performed by Smith, Hitchens and Davies (1982) and covered 117 industrial groupings. Once more they found that productivity within the United States was superior to that of the British industries. However, they also highlighted that there were 24 industries in which the positive productivity gap in the United States was less than twice the British level. Of these 24 industries 'three were in the food sector and eleven in paper, mineral products, textiles and clothing'. Thus they concluded that 'one might characterise these better British performers as low in technological intensity and as low in value added' (Hitchens, Wagner and

Birnie, 1990:7). This suggests that those British industries which were more labour intensive, had a competitive advantage over the more technologically advanced sectors.

The consistently poorer productivity of British firms has increasingly moved the aim of research away from measuring relative productivity, to explaining inferior British productivity. Disparities in the composition of capital and its utilization, has been one of the factors used to explain productivity disparities (Smith, Hitchens and Davies, 1982). Indeed, Prais (1986) discovered in an international comparison of productivity within the metal working industry that despite higher capital intensity within British firms, they displayed lower levels of productivity than their American counterparts.

Labour force characteristics are another factor which have been used to explain productivity disparities. In particular, the educational and skill levels of workers have been identified as central to relative performance, as measured by productivity (Prais, 1981; Prais and Wagner,1983; Daly, Hitchens and Wagner, 1985; Steedman and Wagner, 1987; Prais and Wagner, 1988; Steedman and Wagner, 1989; Prais, 1989; Mason et al., 1994). Previous research into the comparative educational standards of the British and American labour force suggests that again the educational relationship of British workers relative to their American counterparts is one of inferiority (Finegold and Soskice, 1988; Lawrence, 1980).

While comparisons of the organisational and strategic forms of production within the firms help to explain disparities in performance, exogenous factors within the operating environment of the firms are also important determinants of relative competitiveness. For example, institutional factors such as the culture of a region, the provision of venture capital and the nature of government policy are all vital in determining the competitiveness of firms. In undertaking a matched plant analysis it is essential therefore that some commonality exists between the external environments within which the matched plants are operating. With the formation of the European Union and the establishment of a Single European Market, it is likely that this will lead to greater research on a European level. Caution should however be exercized, as the accentuation of cores and peripheries within the European Union may limit the commonality in the operating environments between firms within different parts of Europe and therefore constrain the lessons which can be learnt from undertaking matched plant comparisons.

The National Institute of Economic and Social Research have pioneered the use of the matched plant methodology within a European context. For example, Daly, Hitchens and Wagner (1985) studied 30 engineering firms, manufacturing component parts such as hydraulic valves, screws, and springs etc. in Great Britain and Germany. Steedman and Wagner (1987, 1989) also examined 10 matched plants in Great Britain and Germany, manufacturing kitchen furniture and a further 10 manufacturing ladies clothing. Jarvis and

Prais (1989) also examined a number of retailing outlets in Great Britain and France with Prais, Jarvis and Wagner (1989) focusing upon 14 hotels in Great Britain and 24 hotels in Germany. Further, Mason, Van Ark and Wagner (1993) matched 29 biscuit manufacturers in Great Britain, Germany, the Netherlands and France.

To date only two matched plant studies have been undertaken which focus upon the Northern Ireland economy. Firstly, Hitchens, Wagner and Birnie (1990) carried out a matched plant study of 87 plants, of which they obtained 42 'matches' covering over 32 product groups. Secondly, Roper and Hofmann (1993) undertook a matched plant analysis of 20 firms (10 matched plants). Of the 10 matched plants, 3 were in the clothing sector, 2 were in plastics, 3 were in metal goods, and 2 in the manufacture of electronic components.

With respect to the strengths and weaknesses of the methodology Roper and Hofmann (1993) acknowledge the limitations of the matched plant method where it is characterized by a low sample population and they concluded that for their own sample, 'the small number of Matched plants considered here cannot hope to be fully representative of the four industrial sectors from which they are drawn' (Roper and Hofmann, 1993:27).

The matched plant methodology does however have a number of strengths in ascertaining the micro-level characteristics of regional economies and it is for this reason that the methodology was selected for the investigation of Northern Ireland and Massachusetts small manufacturing firms. For example, Roper and Hofmann (1993:11) state that, 'The strength of the matched plant methodology is that it yields detailed insights into the workings of companies manufacturing similar products in different labour and product market settings'. Additionally, Clulow and Teague (1993:84) contend that, 'Unquestionably, the Organizational Approach has its merits. By identifying and then examining the characteristics of firms in low-productivity regions compared with high-productivity regions, problems and weaknesses can be identified'. The matched plant methodology aims to extend beyond the mere identification of relative performance of firms and highlight those areas which are constraining growth and profitability within the low-productivity region.

Clulow and Teague (1993) also stress some of the problems which are associated with the matched plant methodology. They argue that because of the high dependency upon empiricism, quite often the symptoms of uncompetitiveness are accentuated as opposed to the causes. For example, where previous matched plant analyses within Northern Ireland have pointed to the need to redirect government assistance and upgrade the skills of the labour force, Clulow and Teague argue that this alone is inadequate and instead the restructuring of the governance structure of the economy is a prerequisite to a change in relative competitiveness. Yet, while Clulow and Teague (1993) point to this dichotomy between symptoms and causes, it is also plausible that where the symptoms of uncompetitiveness are not acknowledged, then it is impossible to identify the causes. The matched plant methodology therefore provides a framework wherein the inherent strengths and weakness of firms can be identified, with the importance of each of these factors gauged through the relative performance of the firms within the high-productivity and low-productivity regions.

The matched plant methodology in Northern Ireland and Massachusetts

In applying the matched plant methodology to Northern Ireland firms care was taken to minimize sampling errors. Indeed, where criticism has been directed at the matched plant methodology in the past, attempts were made to overcome these limitations. In particular, care was taken in determining the size of the sample; the sectors which were selected for analysis; the comparative region to Northern Ireland; and the internal characteristics of the firms.

In examining the predominant influences at work within both the matched plants and the different industrial sectors, a sample size was used which was slightly larger than previous matched plant investigations. In total, 100 firms were interviewed, with this being a significantly larger sample than that used by the National Institute of Economic and Social Research; Hitchens, Wagner and Birnie; or Roper and Hofmann. A sample of 100 firms does not negate the criticism of the methodology being unrepresentative of the population, yet, it reduces the likelihood of disparities between the firms being attributable to individual characteristics.

While comparative studies at the beginning and middle of the 20th Century focused upon a wide range of industrial sectors most of the comparative analyses during the 1980s concentrated on a smaller number of industrial sectors. Yet, despite this narrowing of focus, the small size of the samples again questioned the representativeness of the findings to the industrial sectors. The Northern Ireland and Massachusetts matched plant study limited the analysis to the engineering sectors of mechanical engineering, metal products and electrical and electronic engineering firms, and the food processing sector. By limiting the analysis to these industrial sectors and applying a larger sample size than normal, a more representative picture of the comparative performance of Northern Ireland to Massachusetts firms was obtained.

Studies such as those by Steedman and Wagner (1987, 1989) in examining labour intensive industries have neglected to investigate the findings by Smith, Hitchens and Davies (1982) which suggested that within Great Britain capital intensive industries performed relatively worse than labour intensive industries. To examine this perceived disparity in performance the food processing sector was selected as a labour intensive industry with the engineering sectors representing capital intensive industries. Whereas the food processing sector is predominantly labour intensive with low educational and skill levels being required from the workers, the engineering sectors are increasingly capital intensive with worker skills becoming more complex as more advanced technology is introduced. It would be anticipated therefore that the competitiveness gap between the Northern Ireland and Massachusetts firms would be larger for the engineering firms than the food processing firms assuming the validity of the former contention.

In addition, the emphasis upon the emergence of a new technological paradigm (Piore and Sabel, 1984) would suggest that those industrial sectors which are capital intensive would benefit more from the introduction of technologically advanced machinery than labour intensive industries. It would be expected therefore that, assuming the introduction of technologically advanced machinery, firms within the engineering sectors would be outperforming the food processing firms. In addition, Atkinson, in his model of the flexible firm, based the changing form of labour relations upon 72 firms in the engineering, food and drink, retail and financial services sectors. In selecting the engineering and food processing sectors within the Northern Ireland and Massachusetts analysis, it would be expected that if there was a move towards increased labour flexibility within the economies, then this would be identified within the matched plant sample.

Previous international matched plant comparisons have been justly criticized by researchers such as Clulow and Teague (1993) for their identification of the symptoms of uncompetitiveness as opposed to the causes. For example, the majority of the matched plant studies during the 1980s and early 1990s which have compared Great Britain or Northern Ireland with Germany have selected the regions on the basis of one being a high productivity region and the other being a low productivity region. While these studies have succeeded in highlighting weaknesses within the Great Britain or Northern Ireland firms relative to their German counterparts, they have often ignored the importance of controlling, or accounting for, differences in the operating environments of the firms. Without acknowledging the differences in the regional environments, the reasons as proposed for the lower productivity merely reflect the symptoms as opposed to the causes.

Massachusetts was therefore selected strategically as a comparator region to Northern Ireland, displaying similarity on a number of criteria. Firstly, its share of employment within the manufacturing sector was largely similar to that in Northern Ireland, at 18 per cent. Similarly, both regions had experienced a comparable decline in the share of manufacturing as a proportion of total employment since the post war years, with this being replaced within Massachusetts by the service sector, and by the public sector in Northern Ireland. Secondly, both Northern Ireland and Massachusetts have been affected by deindustrialisation of traditional industries in the post-war years. Thirdly, both economies are characterized by social inequality as demonstrated most markedly through spatial variations in unemployment.

While these similarities provide a basis upon which comparative analysis can occur, dissimilarities in the regional environments enable instructive comparisons to be made. These dissimilarities take many forms yet include such aspects as institutional arrangements, an example being the development of private bodies such as the Association of Industrial Manufacturers (AIM) and the Massachusetts High Technology Council (MHTC) which consist of local business men, and is influential in the formulation of political decisions due to their threat to relocate out of Massachusetts. Additionally, government agencies within Massachusetts act in a less interventionist manner than those within Northern Ireland, having fewer schemes of direct assistance to firms. Finally, there is a cultural difference between the regions, which due to its intangibility is extremely difficult to quantify, yet is characterized by aggressiveness within Massachusetts and complacency within Northern Ireland.

The final aspect which was considered important in formulating the matched plant analysis between Northern Ireland and Massachusetts related to the internal characteristics of the firms. As the name of the methodology implies, plants or firms are selected if they fulfill certain criteria yet, previous matched plant studies have often restricted the criteria by which firms are matched. For example Roper and Hofmann (1993) examined firms within four industrial sectors, namely clothing, plastics, metal goods and electronic components. These sectors were selected as a result of their overall market performance relative to that within Germany and thus the taxonomic grouping of firms based upon internal data such as employment, age of firms etc. was ignored. For example, Roper and Hofmann (1993:28) acknowledge that the 'Census of Production information for the UK suggests that large firms typically have higher value-added per employee than small companies', yet, irrespective of this they compare German and Northern Ireland firms which have significantly different numbers of employees. Within the metal working sector the average Northern Ireland firm had 54 employees as compared to 134 employees within their German matches. Northern Ireland firms manufacturing electronic components, on average employed 114, with the reciprocal figure in Germany being 207. Similarly, for clothing and plastics, the Northern Ireland employment levels were 139 and 158 respectively while the German firms employed 216 and 284 workers.

In selecting the matched plants it was imperative therefore that 'like' was compared with 'like' both in terms of the products being manufactured and the levels of employment. Hence, where differences were apparent within the internal functioning of the firm these reflected variations in the production policies being pursued and were not attributable to non-comparative data. It is only when accurate comparisons are made that transferable lessons can be identified between the matched plants and therefore the matching of firms was carefully undertaken.

In addition, it was decided that emphasis should be focused upon the small firm sector, with matched plants having 50 or fewer employees. Particularly within Northern Ireland, the small firm sector has been the primary generator of employment during the late 1980s and into the 1990s. Government policy has targeted the small indigenous sector within Northern Ireland in an attempt to stimulate entrepreneurship and the growth of small firms. For peripheral regions such as Northern Ireland, a poor resource environment has limited the attractiveness of the region to inward investors. This, therefore, places increased emphasis upon the indigenous sector in stimulating regional development. This matched plant research in acknowledging the centrality of small firms to regional development within Northern Ireland ascertains the relative strengths and weaknesses of these small indigenous firms as compared to similar firms within a high-productivity, or core, area. In an increasingly global market it is only through this mechanism that the relative competitiveness of small firms can be gauged and the plausibility of corporate flexibility as an initiator of growth evaluated.

In summary therefore, the matched plants methodology was developed with great care to minimize and eliminate the inadequacies of previous matched plant analyses. Hence, a larger sample size than what had previously been used was implemented along with narrowing the range of study, from a variety of industrial sectors to only two economic groups. In addition, in an attempt to ascertain the extent to which flexible technologies and subsequently flexible working practices had been introduced within the firms - as well as testing the contention that there was a larger gap in competitiveness between Northern Ireland and American capital intensive industries than found within the labour intensive industries - a capital intensive industry and a labour intensive industry were selected for analysis. The sectors of analysis were therefore the engineering and food processing sectors.

Sampling techniques

For firms to be included in the analysis they had to fulfil four basic criteria. The first criteria related to the size of the firms. The research was targeted at the small firm sector and therefore the classification of firms as small was based upon employment criteria, and conformed with the definition of small firms as proposed by the government agency with responsibility for small firms in Northern Ireland, the Local Enterprise and Development Unit (LEDU) in Northern Ireland. Small firms were therefore defined as those firms with 50 or fewer employees at the time of interview. In addition to ensure accuracy of comparison, the matched plants had to display similar levels of employment with deviations in employment between firms only permitted to two employees. The second selection criteria followed the guidelines of the Bolton Report

(1971) in necessitating that small firms must be legally independent of larger corporations or conglomerates.

The third criteria related to the products being manufactured. As already outlined, the research sought to examine a capital intensive and a labour intensive industrial sector, namely engineering and food processing respectively. Problems arose however in terms of the industrial classification of the manufacturing sector. For example, even within the European Union a number of different industrial classification systems exist, although the realignment of each of the systems onto that of NACE (*Nomenclature d'Activité Commerciale Européene*) is planned by 1999. Indeed, Bannock and Daly (1994:98) reiterate this point by stating that, 'there is no international agreement on the break-points or class intervals used in published official statistics'.

In selecting the firms for analysis, it was found that while industrial classifications were provided for all of the Massachusetts firms, the Northern Ireland firms were merely classified alphabetically within broad areas of manufacturing activity. In addition, within the Northern Ireland directory of firms, no distinction was made between firms which manufactured products, provided a service or acted as distributors. In contrast, the Massachusetts trade directories of firms provided a comprehensive listing of information including the spatial distribution of firms, the grouping of firms according to their industrial classification and the categorisation of firms in accordance with their products.

Three industrial classifications were selected for the engineering sector and one for food processing. The United States Standard Industrial Classification (SIC) was used. This system differentiates the activities of firms into a number of divisions, such as construction, manufacturing and services, with each consisting of two-digits. From the two-digit classification firms are then further broken down into industry groups which consist of three-digits, with this finally being disaggregated into four-digit product groups.

For the purposes of the matched plant analysis, firms within the SIC's of 34 (metal goods), 35 (mechanical engineering), 36 (electrical and electronics engineering firms) and 20 (food processing) were selected. In accordance with SIC definitions by the DET (1989), the characteristics of each of these industrial groups were as follows:

> *SIC 34* - Those establishments engaged in fabricating ferrous and nonferrous metal products, such as metal cans, tinware, handtools, cutlery, general hardware, non-electronic heating apparatus, fabricated structural metal products, metal forgings, metal stamping, ordinance (except vehicle and guided missiles), and a variety of metal and wire products.

SIC 35 - Those establishments manufacturing industrial and commercial machinery, equipment and computers. Included within this are the manufacturing of engines and turbines; farm and garden machinery; construction mining and oil field machinery; elevators and conveying equipment; hoists, cranes, monorails and industrial trucks and tractors; metalworking machinery, special industry machinery; general industrial machinery; computers and peripheral equipment and office machinery; and refrigeration and service industry machinery. Machines powered by built-in or detachable motors are also included within this group.

SIC 36 - Establishments engaged in manufacturing machinery, apparatus, and supplies for the generation, storage, transmission, transformation, and utilization of electrical energy. Included are the manufacturing of electricity distribution equipment; electrical industrial apparatus; household appliances; electrical lighting and wiring equipment; radio and television receiving equipment; communications equipment; electronic components and accessories; and other electrical equipment and supplies.

SIC 20 - Establishments which manufacture or process foods and beverages for human consumption. These establishments also manufacture related products such as manufactured ice, chewing gum, vegetable and animal fats and oils and prepared feeds for animals and fowl.

Firms were matched as closely as possible not only in terms of their classification but also by their products to minimize the sampling error associated with mis-matches.

The final criteria by which the matched plants were selected related to the age of the firms. As LEDU has outlined (Buckland, 1993), firms progress through a number of stages, from that of start-up, growth and established to a mature phase. While it is not necessary for all firms to pass through each of these stages of development with some firms remaining within one phase for an elongated period of time, there is an overall correlation between time and development. For example, in the early stages of setting up a business the need for employees will be minimal with both sales volume and turnover being low. Yet, as the firm becomes established it may grow very rapidly in financial and employment terms until it reaches an optimum level of production, beyond which new plant and specialist skills would have to be employed. The firm may stay at this optimum level of production indefinitely, or for as long as the market allows it, unless the firm introduces new products through which the growth phase may be re-initiated.

In accounting for how age may affect growth, the firms were therefore matched by date of birth. While it was acknowledged that this would not necessarily lead to the matching of firms at the same stage of growth, it did increase the accuracy of the matching. In the matching of firms it was realized that precise matching of dates of birth would be impossible, the ages of the matched plants were only allowed to deviate by two years. However, for those firms established during the 1980s it was necessary for the Matched plants to be 'born' within the same year as it is these firms which would be most likely to be passing through the growth phase.

Matching of the firms within Northern Ireland and Massachusetts was therefore tightly controlled. For a match to occur, firms had to employ a similar number of employees (the upper limit being 50 employees); be legally independent of larger corporations; be classified within similar industrial categories and manufacture similar products; and have similar dates of birth.

Characteristics of the sample

In total 100 small manufacturing firms were matched with interviews being performed with the managing directors or owners of each company. The total sample population was split with 70 engineering firms and 30 food processing firms. The engineering firms were further subdivided into industrial classifications with 26 firms being found within the metal goods sector (SIC 34), 22 within the mechanical engineering sector (SIC 35) and a further 22 within the electrical and electronics sector (SIC 36). This subdivision within the engineering sector was included to ascertain the variations within production and performance with respect to various forms of engineering. For example, the nature of production within firms manufacturing electrical and electronic components and products varies significantly from that of fabrication and other structural metal work. Therefore, through the analysis of a cross-section of engineering firms it was possible to assess the differential in performance and flexibility between capital intensive industries and more labour intensive industries in Northern Ireland and Massachusetts.

Within the metal goods sector, the 13 matched plants manufactured a range of products ranging from general metalwork, sheet metal products, wires, springs, aluminium gates, rails and doors to stainless steel fabrications and hydraulic couplings. The products manufactured by firms within the mechanical engineering sector ranged from moulds and machinery for the cement industry, agricultural machinery, textile machinery, power handling and food processing machinery to passenger goods and service lifts, alarms and pressure gauges. Firms within the electrical and electronics goods sector manufactured control panels, printed circuit boards, capacitors and high voltage rectifiers, electrical switchgear and printed circuit boards. Firms within the food processing sector manufactured or processed a range of products, including pies, pizza, coleslaw, sauces, oils, bread and bakery products, potato chips, popcorn, cheese corn, sausages, dog food, fresh and frozen fish, processed wheat flour and sweets and other confectionery. In total 15 food processing matched plants were interviewed within Northern Ireland and Massachusetts.

The distribution of matched plants by employment is demonstrated in Table 3.1.

Table 3.1
The employment distribution of the matched plants

Employment	Engineering	Food Processing
	Number of matched plants	Number of matched plants
< 10	3	0
10 - 20	9	2
21 - 30	12	4
31 - 40	5	5
41 - 50	6	4
Total	35	15

Source: Matched plant analysis

While it was acknowledged that absolute comparability in terms of employment levels would be impossible to attain, this did not diminish the realization that due to the relatively small size of the firms being studied it was critical that employment differentials between the matched plants were minimized. Hence, it was decided that should employment levels within the matched plants vary by more than two employees the match would be rendered invalid. On average, the engineering firms employed 26 workers as compared to 33 within the food processing firms. This is indicative of the tendency for small food processing firms to be more labour intensive than engineering firms.

In terms of the age of the matched plants (Table 3.2), most of the engineering firms were established post-1950. It would be anticipated that the majority of start-ups which were established prior to this date would, assuming survival, have passed through development and growth phases and have employment levels beyond the small firm limit of 50. This therefore accounts for the bias of engineering firms established since 1950.

While it would have been desirable to obtain a cross section of food processing firms across each of the age categories, this was not possible.

Within both regions there were significantly fewer food processing firms than engineering firms and therefore the population was smaller. This explains the inability to sample firms established between 1930 and 1949 and 1980 to 1989. It should be noted, however, that this does not invalidate the results as the firms which were matched had similar dates of birth.

Table 3.2
The age distribution of the matched plants

Year	Engineering	Food Processing
	Number of matched plants	Number of matched plants
1900 - 1929	2	5
1930 - 1949	3	0
1950 - 1969	8	5
1970 - 1979	13	5
1980 - 1989	9	0
Total	35	15

Source: Matched plant analysis

The matching of small firms was therefore performed as accurately as possible. Indeed not only did the matched plants satisfy the criteria of being legally independent of larger firms or conglomerates, but they also displayed similarity with respect to the products being manufactured, the age of the firms and the employment levels. Further, while it was acknowledged that the matched plant methodology does not seek to be representative of the population, a representative sample within Northern Ireland - in terms of the industrial concentration of engineering and food processing firms in the industrial structure as well as the geographical distribution of firms within the province - was established in an attempt to increase the probability that the characteristics of the small firms were typical of similar small firms within Northern Ireland.

Matched plant questionnaire

The questionnaire which was used in the analysis consisted of a number of sections, each to determine the underlying organisational structure within the firms. The questionnaire consisted of nine separate sections and each of these is outlined briefly.

Section one - Company background and average scale of production:

This section sought to outline the primary characteristics of the firms and therefore ensure the validity of the match. Fundamental questions were asked pertaining to the identification of the firms in terms of their names and addresses, with the industrial classifications of the firms also being recorded. This section also ensured the validity of the match by asking when the firm was established. This was important because on a number of occasions the information within the trade directories - especially that for Northern Ireland - was inaccurate. The first half of this section therefore identified the firm being interviewed in terms of each of their taxonomic groups, that is their industrial classification, the products being manufactured, and their date of birth.

The second part of this section ascertained the financial turnover of the firms both during the financial year prior to the interview (which was 1993) and five years previous, namely 1988. As was aforementioned turnover was not an ideal measure of the performance of the firm for a variety of reasons. For example, international comparisons mean that there is a discrepancy between the producer and consumer prices of goods within each of the regions and therefore these factors had to be accounted for in the comparisons. In addition, turnover merely reflects the value of goods leaving the plant with no account being made for value added, or the value of inputs. Therefore, where the tendency was found amongst Northern Ireland firms to purchase a higher percentage of inputs than Massachusetts firms, with these being only slightly modified during production, then turnover values although appearing similar were not representative of either the profit levels or the value added through production.

However, despite the limitations of information on financial turnover, it was still used for two reasons. Firstly, through the pilot interviews it was evident that the managers were unwilling to provide more precise financial information relating to net profit and cost of sales, and as many of these small firms merely submitted abbreviated accounts to the companies registry, this information could not be obtained through published sources. The second reason why information on financial turnover was used related to the collection of information for 1993 and 1988. The firms were asked in section two, if they were producing new products in 1993 compared to 1988, further in section nine the age and technological composition of their machinery was examined. From these two questions it was apparent that the majority of firms both within Northern Ireland and Massachusetts in 1993 were producing a similar range of products to that of 1988. In addition, the proportion of machinery under the age of five years within both regions was low suggesting that the composition of production had not changed significantly during this period. Therefore, in assuming that the proportion of the final output manufactured within the firms had not changed dramatically over the five year period, then the change in financial turnover could be viewed as an estimation of the growth of the firms, after allowance was made for inflation and price changes.

The final part of the first section determined whether or not the firm had a written business plan and the primary objectives of such.

Section two - Markets:

This section ascertained the competitive market within which the firms were operating. Firms were asked to outline the number of direct competitors that they had and their comparative advantages and disadvantages relative to them. In addition, firms were also asked to identify the market distribution of both their inputs and outputs and the incidence of subcontracting.

These questions highlighted the market sphere within which the firms operated. For example, while the Northern Ireland firms identified a limited number of direct competitors, most of their competitors were located within the Northern Ireland region. In addition, a high proportion of outputs was sold or distributed within the local market area of Northern Ireland. This illustrates that the market area of Northern Ireland firms is largely internal with managers perceiving only those firms within Northern Ireland that produced the same products, as representing their main competitors. Conversely, within Massachusetts, the firms often identified a larger number of direct rivals than was apparent within the state trade directories. In addition, these Massachusetts firms also distributed a greater proportion of their output beyond the local state as compared to Northern Ireland, a factor that reflected their view of the competitive environment. This factor is important to understand because the nature of the competitive environment within which the firms perceive themselves to be operating, will determine the aggressiveness of their strategic policy and the business objectives of the firms.

Section three - Manpower:

This section determined the employment structure of the firms and how this had changed since 1988. Information was attained for the employment level in 1988 and the manner by which the incidence of part-time and temporary workers had changed over the 1988 to 1993 period.

The hierarchical breakdown of the employees within the firms was also obtained. Within the small firms however there was a greater overlap of functions, than would have normally been found within a larger and more structured firm, with this in some cases culminating in the inability of managers to formally and accurately categorize the workers. While attempts were made in this section to acquire comprehensive data on the remuneration levels of employees, this question had a low response rate.

Worker relations were also examined within this section. For example, it was determined whether or not employees belonged to a union or other form of collective organisation and the manner by which remuneration was determined. Further, the nature of management/worker relations was examined, with managers being asked to state if there had been disputes or strikes within the firms and the number of employees that had left the firms in the 12 months prior to the interviews, along with their reasons for leaving.

While these questions in themselves were significant in terms of small firm dynamics as compared to larger enterprises, and the implications of such for employee protection and job security, they were also important within a regional context. For example, was there a higher level of numerical flexibility amongst workers within one of the regions and was this accompanied by a diminuation of worker rights where greater instability in the external labour market was found. Further, by linking these findings to those of the performance of firms through changes in employment and turnover, then the efficacy of these policies could be assessed.

Section four - Shop-floor skills:

This section established the incidence of formal qualifications amongst the work forces of each of the firms. In addition managers were asked to highlight the efficiency of the shop floor workers and their ability to overcome problems which they encountered during the production process.

The aim of this section was to identify the extent of formal academic and vocational qualifications upon which training was undertaken. Further, the managers were asked if they believed that their workers had the skills and training which were necessary to perform their jobs efficiently. The response to this could then be examined within the context of the provision of training being provided by the firm, as highlighted within section five.

Section five - Training of the work force:

Having outlined within section four, the standard of formal educational and vocational qualifications amongst the workers this section ascertained the extent to which the work force had received training in the 12 months prior to the interview. This training included informal on-the-job training and the implementation of seminars and workshops on-site, to college day release and night courses and conferences. In addition, the interviewees was asked to estimate the cost of training schemes and whether the firm had financed the training in full, or if the employees had contributed.

Information was also obtained on the extent of public/private contact between the firms and government agencies during the previous 12 months and the form which this had taken. Managers were also asked to identify the various forms of assistance which they had received previously from such agencies and to state if they believed that additional financial assistance would increase the scale of training within their firms. The objective of these questions was to ascertain the level of financial dependency upon government agencies as well as investigating the effectiveness of such agencies in increasing the awareness of firms in terms of the available schemes to assist in the training of employees.

The final part of this section investigated the extent of job-sharing and multi-skilling within the firms. Employers were asked about the extent of demarcation within the firms and the level of employee interaction during the production process. These questions therefore assisted in outlining the manifestation of functional flexibility within the production process with higher levels of interaction and the elimination of lines of demarcation being indicative of both the higher skill capacity and multi-skilling of the work force.

Section six - Supervisors:

This section in conjunction with section seven examined the hierarchical structure of management within the firm. Supervisors are regarded as middle-managers and within the Fordist mode of production they strictly monitored the performance of other shop floor workers in terms of the speed at which they worked and the quality of the work which they produced.

Within the new model of flexibility, the role of supervisors is radically altered. For example, one of the characteristics of flexibility is that there will be a diminuation in the visibility of supervisors within the firm with them being absorbed into the shop floor work force. This is achieved through the multi-skilling of the work force as a whole whereby, through increasing their skills base, employees will be capable of a wider range of tasks. In addition, workers will have the necessary skills to identify errors in production at an early stage (as opposed to errors being detected by a quality control division within the firm) and either adapt production to eliminate problems within the production or rework the defective parts and reduce total waste.

This section therefore examined the incidence of supervisors within the firms and the number of employees for whom they had responsibility. In addition, where supervisors were found, their role within the firm was examined in terms of the major functions which they performed and their ability to execute higher order functions such as planning, design, marketing, purchasing and R&D. This was asked so that an erroneous assumption could be avoided, namely, that where supervisors were apparent this was indicative of Fordist methods of production and where they were absent that flexibility had been adopted. This assumption highlights the inherent ambiguities with the term flexibility. For example, does the elimination of a middle-management level within the hierarchy of control assist in the establishment of flexibility of production or flexibility of managerial control? While it could be argued that reducing the number of supervisors is a direct response to attaining greater flexibility amongst the shop floor and technical workers with a consequent reduction in the level of control needed over these workers, it could also be argued that maintaining a supervisory level is in line with the attainment of flexibility, *providing* that this middle-management strata is competently skilled.

Therefore, should a middle-management level be absent within the firm this could be as a consequence of the multi-skilling of the work force as a whole. Alternatively, where supervisors are found within the firm, this may not be an indication of the necessity to continually monitor and check the performance of other workers, but rather to accentuate the level of flexibility within the firm through a high skill and competence level while maintaining a structure of hierarchical control. It should be noted that flexibility *can* occur within the production process, within the framework of a hierarchical structure. Yet, while theoretically, the attainment of a flexible work force should reduce the need to continually monitor and check the work of employees, the existence of supervisors and production flexibility are not mutually exclusive providing that the supervisors are highly skilled and competent in a range of tasks throughout the firm. In response to this, the questionnaire established the educational and skill levels of supervisors along with the nature of the training which they received in conjunction with the role which they played within the firm and their capacity to perform higher order functions.

Section seven - Managers:

The managerial section of the questionnaire highlighted a number of factors. Firstly, the number of managers was identified and the formal qualifications held by such, including the date of attainment. This section followed the previous three sections in establishing not only the standard of formal educational and vocational skill levels, but also the extent of training which the managers had received and their ability to perform certain functions within the firm. The extent of overlapping of functions was also examined to ascertain the flexibility of managers in performing a range of tasks. Caution had to be exercized here because in many cases where one manager performed a variety of tasks this was merely a function of the size of the firm and the human resource limitations associated with a small enterprise.

The flexibility of managers was also examined in terms of their ability to perform shop floor tasks and the ease with which these were carried out. Managers were also asked whether or not they trained shop floor workers and supervisors, and where this was found to occur it was indicative of the managers being capable of performing shop floor functions efficiently.

The final part of this section examined the perceived strengths and weaknesses of managers in running the firms. In addition, managers were asked to highlight how managerial training contributed to the overall performance of the firm. Through this question the attitudes of managers were explicit not only with respect to the importance of training but also the manner in which they believed training could assist in the development of the firm.

Section eight - Specification of the product:

This section sought to identify some of the characteristics of the products produced. For example, information was obtained pertaining to the reject rate of production and the manner by which quality was incorporated within the production process. These two questions assisted in the recognition of Quality Control Circles (QCC's) and the effects of such upon the quality of products. Within the flexible production system firms are expected to reduce costs through an improvement in quality. This is attainable through the introduction of work teams coupled with the reduction in lines of demarcation and the greater interaction of employees. These two questions therefore tested the viability of this hypothesis and established if the incorporation of QCC's did lead to reductions in the reject rate of production.

This section also examined whether or not the firms manufactured for stock and if they undertook R&D. While it is proposed that it is only through the undertaking of R&D that a firm can grow rapidly and that within the new model of flexible production R&D will become an integral part of the activities of firms, the reality of this within the sample was assessed.

Section nine - Production:

The concluding section ascertained the technological environment within which the firms were working. For example, information relating to the age of machinery was derived along with that relating to the incidence of technologically advanced machinery such as CNC's, CAD, CAM, FMS etc. In addition, the managers were asked to highlight the extent of interaction which they had with their machinery suppliers. Managers were also asked if the employees within the firms adapted the machinery, or if they were involved in day-to-day general maintenance. It has been illustrated in previous matched plant comparisons that where preventative maintenance was undertaken this had a positive effect upon the operating of machinery by reducing the number of breakdowns (Hitchens, Wagner and Birnie, 1990). Therefore these questions were designed to establish if there was a direct correlation between the extent to which employees adapted the machines and the number of breakdowns.

Although each of the sections of the questionnaire are distinct through the specific information which they create none of the sections are mutually exclusive. Instead, there is an explicit inter-twining of all of the variables. For example, in the same way as it is fallacious to study the external competitive environment of the firm to the exclusion of the internal forces at work within the small enterprise, and vice versa, it is also negligent to study specific aspects of the internal workings of the firm to the omission of others. Findings from each of the sections are therefore examined within the overall context of the firm and the nature of corporate flexibility determined.

4 Comparative analysis of matched plant growth: Northern Ireland and Massachusetts

> It can be said without exaggeration that in the long-run probably nothing is as important for economic welfare as the rate of productivity growth *(Baumol et al, 1989: 1)*

Introduction

The attainment of 'growth' has occupied a central position in the quest for economic prosperity, both on a macro- and a micro-economic scale. It is only through the attainment of a competitive advantage that growth can occur whether this be with respect to the economic performance of regions or individual firms. Carr (1990:13) states that;

'Competitiveness broadly refers to the ability of a business...to sustain future growth...relative to actual or likely competitors... [The] concept of sustainability is important because it is this that rules out superficial analysis and forces deeper examination of fundamental factors underlying the strength or weakness of an economy or of a particular business'.

Therefore, by nurturing and consolidating a competitive position the future trajectory of a region or business will be determined. At each stage of development, it is only through the embracing of competitiveness that growth will occur. Yet, despite various forms of analysis to identify the varying levels of competitiveness at a micro-level, it is competitiveness at the national level which has gained the most attention. The IMD (1991:8) has defined the competitiveness of a country as, 'The extent to which a national environment is conducive or detrimental to business'.

Clearly, such a definition fails to identify the precise nature of national competitiveness. However, within the UK Governmental White Paper on Competitiveness (HMSO: 1994), the Ministerial committee with responsibility for the investigation, adopted a more comprehensive definition of competitiveness as used by the OECD, namely that;

'Competitiveness is the degree to which a nation can under free and fair Market conditions, produce goods and services which meet the test of international Markets, while simultaneously maintaining and expanding the real incomes of its people over the long-term.' (Goodhart, 25/5/94:10).

The White Paper acknowledged the dynamism of competitiveness and stressed the necessity for change and innovation if the UK was to consolidate and improve its competitive position relative to other nations.

There are a number of reasons why it is important to ascertain the growth of small firms, yet it is the perceived ability of small firms to generate new employment which has been emphasized most. Small firms have been identified as major providers of employment creation within a number of studies (Storey et al., 1987; Birch, 1979; Smallbone et al., 1993; Woods, Blackburn and Curran, 1993). Yet, it is also emphasized that the ability of small firms to generate employment is confined to those which through the attainment of a competitive position can consolidate their relative market position. While the death rate is disproportionately high amongst small firms those which survive may contribute substantially to employment growth. For example, within a study of new indigenous manufacturing firms established between 1973 and 1986 within Northern Ireland, the Republic of Ireland and Leicestershire (NIERC, 1989) it was found that these firms accounted for 16.5 per cent, 14.5 per cent and 14.7 per cent of manufacturing employment in 1986, respectively. Small firms are therefore an essential component of regional development and for this reason it is essential that their growth capacity is identified along with the constraining mechanisms which may be hindering competitiveness and growth.

Despite claims by Flexible Specialization and Regulation theorists that new forms of flexibility within the workplace will assist in the generation of higher levels of productivity and growth little evidence has been produced to support these claims. It is therefore the aim of this chapter to analyse the validity of these claims and examine to what extent the small engineering and food processing firms within the two regions experienced growth between 1988 and 1993. The examination and determination of growth is based upon the variables of turnover and employment. While it is important to ascertain changes within these variables it is also requisite that such changes be understood within the wider framework of the firms concerned and the internal and external economic forces at work within them[1]. Analysis is therefore made of the mechanisms within the sample which characterize the changes in turnover and employment over the period.

The findings relating to growth provide the foundation upon which growth disparities can be more fully understood in terms of the functioning of the plants. The following questions must therefore by addressed, namely;

1. Is there any evidence of growth within the firms?
2. Is there an identifiable disparity in the extent of growth between the two regions?
3. Is there a difference in the extent of growth between the two sectors of analysis?
4. Where disparities in growth are apparent, what are the contributory factors of such?

Small firm growth

North and Smallbone (1993) carried out a study of 306 firms geographically situated within three varying landscapes. One hundred and twenty six firms were based in London (65 of which were in Inner London), 100 firms were in Hertfordshire and Essex and 80 firms were within areas which had settlement populations of under 10,000 people. Through the utilization of established empirical data and the surveying of each of these firms, they attempted to examine how the growth of the firms changed over a 10 year period and the relationship of such to the creation of jobs. Additionally, they identified geographical variations in the level of job creation and accounted for these differences in terms of internal and external factors at work within the different businesses and environments.

North and Smallbone (1993:4) discovered that 'job creation was closely tied to the growth of the firms in output terms (measured in terms of deflated sales turnover 1979-90) and that nearly all the new jobs were created by firms which at least doubled their output over the decade'. They also found that where turnover increased by 250 per cent this equated to an increase of 200 per cent in employment levels. However, within the printing and electronics manufacturing sector for a 100 per cent increase in employment levels to result, this required a 300 per cent increase in turnover. Of greater importance than the proportions of turnover and employment which they found, are the inherent implications of their work. They make a clear correlation between turnover and employment, suggesting that they are interrelated variables with the increase in one being accompanied by an increase in the other. These findings are also supported by other findings (Roper, McFerran and Hewitt-Dundas, 1996) whereby, examination of 706 small manufacturing firms throughout Ireland found that there was a significant positive correlation between real turnover and employment growth.

These findings have important implications for peripheral economies such as Northern Ireland where high levels of unemployment are endemic to both the economy and social fabric of the region. If the overall performance of industry can improve this will ultimately result, on a national level, in higher levels of GDP and subsequently higher standards of living, along with a reduction in the unemployment rate. In terms of the sample of Northern Ireland and Massachusetts firms, the question must be asked; to what extent is there a positive relationship between turnover and employment growth?

Growth within the matched plants - turnover and employment based measures

Turnover is defined as the total sales revenue of the firm. To facilitate a useful comparison firms were grouped according to their Standard Industrial Classification and analysed within these categories. For each of the firms information was requested relating to the turnover value of the firm during the financial year prior to the interview - which was 1992/93 - and the base year 1987/88. This created a five year interval over which growth could be determined. For each of the sectors average real turnover growth rates were calculated in the following way. The turnover values for the firms within each of the sectors were aggregated for 1988 and 1993. The 1993 values were then converted into 1988 prices[2] and the average real turnover growth rates determined for each of the industrial sectors (Table 4.1).

These values suggested that in real terms, turnover growth within the Massachusetts firms increased by a greater percentage than for the Northern Ireland firms. The mechanical engineering sector is an anomaly to this trend with the Northern Ireland firms within this sector outperforming their Massachusetts counterparts. Generally, however, the results for the sample companies present a clear regional picture in terms of growth of real turnover.

The Massachusetts firms were out-performing their Northern Ireland counterparts as regards every aspect pertaining to the turnover of the firms. While a plausible explanation for this higher level of growth within the Massachusetts firms could be that their sales turnover was lower in 1988 and this provided greater scope for expansion, such an explanation is inaccurate as is manifest by the average adjusted turnover values (Table 4.1). These demonstrate that the average turnover of firms within each of the sectors was higher in Massachusetts in 1988. Indeed the mechanical engineering sector within Massachusetts which experienced negative real turnover growth over the period from 1988 to 1993, had the highest *relative* turnover with the Massachusetts turnover level in 1988 being 142 per cent that of their Northern Ireland counterparts. While the relative turnover of the mechanical engineering firms within Massachusetts declined significantly during the five years, by 1993 the average turnover of Massachusetts firms was still above - equating to 17 per cent - that of the Northern Ireland firms. Therefore, not only was the absolute output per firm higher in Massachusetts but, with the exception of Massachusetts firms within the mechanical engineering sector, their real turnover growth was also significantly greater.

Table 4.1
Percentage change in real turnover and employment, by industrial sector and region

	Average Turnover 1988 (£000s)	Percentage Growth Real Turnover	Percentage Growth Employment	Average Turnover 1993 (£000s)
Metal Products Northern Ireland	1,125	-7.2	14.0	1,044
Metal Products Massachusetts	2,208	20.74	1.4	2,666
Mechanical Engineering Northern Ireland	1,066	23.4	-5.8	1,316
Mechanical Engineering Massachusetts	2,583	-9.78	14.5	2,330
Electronics Northern Ireland	1,522	18.15	7.8	1,798
Electronics Massachusetts	3,361	52.8	-6.0	7,705
Food Processing Northern Ireland	1,160	6.15	13.5	1,301
Food Processing Massachusetts	2,150	48.6	27.5	3,196

Notes: Average Turnover 1988 calculated in £ Sterling for Northern Ireland firms and $ US for Massachusetts firms.

Source: Matched Plant Survey

The work of North and Smallbone (1993) amongst others emphasise the existence of a positive relationship between turnover and employment growth. In light of the significance of this for employment creation within regional economies, and within the context of the findings pertaining to turnover growth, the change in employment was also examined within the matched plant survey.

Each of the firms visited provided information with respect to their employment levels both in 1988 and 1993 and from this the percentage growth in employment between 1988 and 1993 for each of the industrial sectors could be calculated (Table 4.1).

It should be noted that as a result of the firms being matched by employment levels the total number of employees within Northern Ireland approximated that of the Massachusetts firms and therefore accurate comparison between the regions was facilitated.

From the results it is possible to identify those sectors which have experienced higher rates of growth than others[3] . Within both regions firms within the food processing sector experienced the greatest average increase in employment levels, however in percentage terms while the food processing sector remained the most significant industrial sector within Massachusetts - as measured by employment growth - it was placed slightly behind the metal products sector in Northern Ireland. There was a clear distinction between the engineering and food processing sector within Massachusetts with average employment change in the former being either close to zero or negative whilst in the latter employment changed by an average of 7.7 employees per firm, representing a percentage change of 27.5 per cent. These findings would suggest that it is the food processing sectors within both regions (along with the metal products sector within Northern Ireland) which are growing at a faster rate than the other sectors under examination in terms of employment.

Yet why are there differences in the extent to which new employment is generated between each of the industrial sectors? The most obvious reason for variations in levels of growth - as measured by employment values - is the differences in the inherent characteristics of the sectors. For example, as is demonstrated within Chapters Five and Six, the food processing sector is very labour intensive, whereas in other industrial sectors, such as electronics, technological developments and the introduction of numerically controlled and other advanced machinery have facilitated a reduction in the intensity to which labour is incorporated within the production process. The food processing sector within both regions demonstrated the largest increases in the levels of employment with the average food processing firms within Northern Ireland and Massachusetts increasing their work forces by 4.5 and 7.7 employees, respectively. However, the findings pertaining to turnover growth do not support this positive industrial bias in performance towards the food processing sector. The hypothesis of North and Smallbone (1993) concerning a positive relationship between turnover and employment is therefore questioned within the context of this study. It is important therefore to determine to what extent complementarity exists between turnover and employment growth.

The relationship between turnover growth and employment growth

With the exception of the work by North and Smallbone (1993) which argued that there would be a direct correlation between turnover and employment growth there is a dearth in the small firms literature in terms of the analysis of both employment and turnover changes. While numerous studies have examined the employment contribution of cohorts of small firms in terms of survivors and failures, or runners and plodders, (Storey *et al.,* 1987; Rajan and Pearson, 1986; Gallagher and Miller, 1991) little attention has been focused upon the direct relationship between turnover and employment growth. If employment creation is to take a central place in the policy objectives of regional economies then it is paramount that the relationship between small firm employment growth and sales growth are fully understood This section therefore examines the interconnectedness of these two aspects of growth.

Having calculated the average real turnover growth and employment growth levels for each of the industrial sectors within both regions the differential between these values were calculated. In addition, simple correlation coefficients between turnover and employment were determined for each of the sectors (Table 4.2).

From the results it is apparent that the findings for the sample do not conform to those proposed by North and Smallbone (1993), but instead distinct disparities between growth as measured by turnover, and that as measured by employment were apparent. The differentials for the percentage growth in turnover and employment (Table 4.2) are indicative of the mis-match between these two indicators with a large differential suggesting that turnover and employment have followed different growth trajectories over the period. Where previous studies have found a positive relationship between turnover and employment growth, the results demonstrate that for the metal products and mechanical engineering sectors there is a significant negative association as found through the use of the Pearson chi-squared test. Within the electronics and food processing sectors no significant correlation was found between turnover and employment growth. This would suggest that within the sample of matched plants there are factors at work which are counteracting the traditional co-dependency associated with turnover and employment growth.

Both regional and sectoral differences are apparent in terms of turnover and employment growth and the extent of complementarity between these factors. With the exception of mechanical engineering, the percentage change in turnover values for the Massachusetts engineering firms greatly exceeded that for the Northern Ireland firms. Additionally, within the food processing sector, the growth in turnover was accompanied by a substantial growth in employment, this is however more the exception than the rule; as for the remainder of the sectors, the percentage increases in turnover did not correspond to that of employment to the same extent. (As was previously suggested, this is a direct reflection of the higher labour to capital ratio which is requisite within the food processing sector).

Table 4.2
The differential and correlation between the percentage change in turnover and employment growth by industrial group and region

	Metal Products	Mechanical Engineering	Electronics	Food Processing
Northern Ireland	21.2	29.2	10.3	7.4
Massachusetts	19.3	24.2	58.8	21.1
Correlation Co-efficient	-.57 ρ = .024	-.91 ρ = .001	-.25 ρ = .358	.13 ρ = .637

Note: The Pearson Chi-Squared test was used to establish the nature of the correlation.

Source: Matched Plant Survey

The electronics sector demonstrated the largest disparity between the regions not only in terms of turnover and employment growth, but also the disparity between the level of turnover and employment growth from 1988 to 1993, that is, 52.8 per cent and -6 per cent respectively. The electronics sector in particular challenges the traditional conceptualization of turnover and employment growth and compels a deeper examination of the strategic and operational activities of the firms concerned.

Finally, the firms within the food processing sector are distinct from the engineering sector in that there was complementarity between turnover and employment growth. The extent of this growth and the balance between turnover and employment was however different between the two regions. For example within Northern Ireland a 6.15 per cent increase in turnover over the 5 years led to a 13.5 per cent increase in employment. Conversely a 48.6 per cent increase in real turnover within Massachusetts was accompanied by a 27.5 per cent increase in employment. This would therefore suggest that the Massachusetts firms were being more efficient in their utilization of labour and significantly increasing their output per employee over the period as compared to Northern Ireland firms. To examine this proposition the level and change in sales per employee was investigated.

Output per employee growth 1988 to 1993

In examining the hypothesis that the Massachusetts firms deployed their labour resources in a more efficient manner than the Northern Ireland firms, the sales per employee and the growth in sales per employee between 1988 and 1993 were calculated for each of the industrial sectors (Table 4.3).

For each of the sectors output per employee as measured in real terms, was higher in 1993 within the Massachusetts firms. Of particular interest are the findings relating to the mechanical engineering sector. Although the percentage change in output per employee was smaller within the Massachusetts mechanical engineering firms, in absolute terms the value of output per employee was higher for both 1988 and 1993, with the increase over the intervening period also being marginally greater (allowing for currency conversions).

Having outlined the predominant changes in the levels of turnover, employment and output per employee for each of the industrial sectors over the period from 1988 to 1993, it is beneficial to further examine some of the underlying characteristics and mechanisms of the firms which were contributing the patterns of turnover and employment growth.

Table 4.3
Growth in sales per employee from 1988 to 1993 (3)

	Metal Products	Mechanical Engineering	Electronics	Food Processing
Sales Per Employee 1988				
Northern Ireland	58,396	41,290	67,156	34,730
Massachusetts	90,753	93,373	181,137	76,785
Sales Per Employee 1993				
Northern Ireland	47,475	54,118	73,574	34,345
Massachusetts	108,102	98,479	294,466	89,490
Change in Sales Per Employee 1988-1993				
Northern Ireland	-10,920	12,828	6,418	-385
Massachusetts	17,349	5,106	113,329	12,705
Percentage Change in Sales Per Employee 1988-1993				
Northern Ireland	-18.71	31.06	9.55	-1.11
Massachusetts	19.1	5.46	62.56	16.5

Note: Sales per employee and change in sales per employee were measured in real terms with 1993 sales values being adjusted in line with the Producer Price Index. Sales per employee were measured in £ Sterling within Northern Ireland and $ US within Massachusetts.

Source: Matched Plant Survey

Operational mechanisms contributing to growth

While it is important to comprehend the overall change in growth for each of the industrial sectors, these figures alone fail to present a clear picture of the organizational and strategic properties of the firms. To help understand the predominant factors at work within the firms, some of the mechanisms which were contributing to differential changes in turnover and employment growth within the sample were examined.

Restructuring of markets

Within some of the sample firms in Northern Ireland the loss of market share combined with the inability to effectively penetrate new markets had forced the companies to evaluate their competitive position. In some instances a high level of competition within the local market area had led to the reduction in sales (see Case Study One) while in other cases the over-dependence upon subcontracting relationships and the collapse of these markets had led to the re-evaluation of relative market position (Case Study Two).

Case Study One:
A company established in 1960 within Northern Ireland manufactured structural fabrications and sheet metal work. Between 1988 and 1993 turnover declined by 22.9 per cent while employment increased by 9.7 per cent from 41 to 45 employees. The level of sales per employee had dropped considerably over the five year period. This anomaly could be attributed to the manner in which the firm had restructured its market. The firm being located within a highly competitive market identified a fall in turnover and attributed this to their competitive disadvantage relative to their competitors in the provision of consultancy services in conjunction with manufacturing.

The firm therefore attempted to incorporate this aspect within their operations. Due to falling turnover levels, manufacturing operations had to be reduced and skilled engineers and draughtsmen employed at the expense of shop floor workers. The primary objective of the company was to develop the service sector of the business as a means of future growth. Therefore competitiveness would be improved through the vertical expansion of the firm, yet as this expansion was accompanied by a weakening of the manufacturing base the actual effects upon competitiveness is debatable.

Case Study Two:
A Northern Ireland company manufacturing metal fabrications experienced similar problems to that of case study one. Historically, the firm had a strong dependence upon subcontracting relationships with three contracts representing 100 per cent of its sales in 1988 and 6 contracts representing 80 per cent of its sales in 1993. It was the loss of two of the three contracts in 1990 which left the firm vulnerable to market changes. Acknowledging a high level of local competition within this industrial sector, the firm decided to diversify its market through the incorporation of a service element into the business in an attempt to gain a competitive advantage.

During the period of restructuring, the firm shed 60 per cent of its employees. From June 1991 to June 1992 the firm employed a number of technical staff including draughtsmen and mechanical and electrical engineers. By 1993 the numbers employed had increased to 50, only five fewer than in the pre-restructuring period. In addition, turnover had returned to 66 per cent its 1988 value.

These firms were at a competitive disadvantage relative to their competitors due to an inability to provide their customers with consultancy and design services in conjunction with manufacturing. Acknowledging this position the firms rationalized their manufacturing operations through the shedding of un-skilled and semi-skilled labour and employed technicians and draughtsmen. While the restructuring had culminated in a negative level of turnover and employment growth from 1988 to 1993 it is plausible to suggest that the changes in the market orientation of the firm would improve their competitive positions and subsequently lead to a reversal of the prevailing growth trend.

In light of the inefficiencies within the firms in terms of manufacturing (Case studies One and Two) and marketing (Case study Two) any changes in terms of the market orientation will merely act in a substitutive as opposed to a complementary manner relative to the original business. Indeed while the intention of incorporating a consultancy element within the business was to improve the competitive position of the firms relative to their competitors, by reducing the manufacturing component, this merely served to weaken their overall market position.

Restructuring as a business strategy was most commonly found amongst the metal products sector within Northern Ireland. Where this occurred it was accompanied by a decrease in both turnover and employment from 1988 to 1993. The inherent weaknesses within these companies and the debilitating effects of diversifying their market bases leads to questions regarding the competitive legitimacy of these companies in the future.

Non-innovative manufacturing practices

In the case of firms restructuring their markets it was found that both turnover and employment declined from 1988 to 1993. Where no strategic or organizational innovations were apparent in the operating mechanisms of firms over the five year period this also led to a decline in turnover and employment (Case Study Three).

Case Study Three:
A Massachusetts family firm established in 1937 manufacturing agricultural machinery experienced a decline in turnover of 10.8 per cent, and a decline in employment of 9 per cent from 1988 to 1993. The main market for the firms products was a niche market of animal feeders. The firm had manufactured feeders for many years and had no desire to move into other areas of production. Forty per cent of their output was attributed to two firms with which the company had a long-term (informal) relationship.

There were no part-time employees within the firm, and each of the workers had permanent contracts of employment. In addition, all of the shop floor employees had certificate or associate level qualifications and the managing director had a Masters degree in mechanical engineering. Yet, as the main products had not changed during the previous 15 years - with the exception of minor adjustments to the automatic chute feeds and time delayed droppers - this was indicative that the inherent skills of both the workers and the management were not being exploited to the benefit and progression of the firm.

This firm was therefore static in terms of both technological and organizational innovations. The maintenance of market share within a niche market had rendered the firm complacent with regards its competitive position and this was reflected in the identified business objective of the managing director as being to maintain the firm's present market position as opposed to obtaining growth.

Competitiveness is dynamic and while the attainment of a competitive advantage may lead to market growth, unless the mechanisms associated with competitiveness are adopted as integral parts of the business strategy the acquired competitiveness will be counteracted by marginalization, stagnation and decline. The UK White Paper on Competitiveness (HMSO, 1994) reinforced this perception in stating that; 'Competitiveness is dynamic. We cannot ever afford to stand still. For as we change and innovate, the outside world also changes.' This view is frequently echoed in the policies of Development Agencies. For example, the small firms development agency in Northern Ireland, (LEDU, Local Enterprise and Development Unit) consciously provides preferential support to those companies which are characterized as having the potential to grow. Therefore, as is evident within the matched plant study, in the absence of innovative and strategic developments within a firm previous competitive advantages will be compromized and relative performance will be characterized by negative growth and general decline.

Firms with weak market positions

A number of studies have demonstrated that there is a positive relationship between market share and growth. The Profit Impact of Market Strategy (PIMS), (Buzzell & Gale, 1987) demonstrates that the return on investment will be very different for firms with a high as opposed to a low market share. In addition, Storey (1994) concludes that despite the lack of research into the effect of market positioning upon profitability and growth, results in general would suggest that relative market share is a vital component of growth amongst smaller firms.

Small firms by virtue of their size often occupy a significantly small proportion of total market share. This creates vulnerability to price, quality and other market fluctuations unless the firm can develop generic strategies (Porter,

1980) which are not easily matched by its competitors. Within the matched plant study, many of the small firms were characterized as having a weak market share (Case Studies Four and Five) as a result of two factors. Firstly there was a high level of competition and secondly, this competitive environment was reinforced by the over-concentration of output within the local market (see Chapter Seven). Firms responded to this situation by either protecting their markets through the attainment of niches and/or the establishment of subcontractual relationships (Case Study Four), or by proactively seeking to improve their competitive position (Case Study Five).

Case Study Four:
A firm manufacturing sweets and other sugar confectionery products was established within Northern Ireland in 1972. Turnover within the firm increased by 2.7 per cent from 1988 to 1993 and this was accompanied by an increase in employment of 23.3 per cent. The firm stated that it competed directly with 6 other manufacturers and 95 per cent of its output remained within the local market. There was a strong dependence upon one subcontracting relationship which represented 60 per cent of total sales thus creating a high level of vulnerability to future changes in the market.

The management stated that because of the relatively high dependence upon this one buyer the plant was frequently inspected by management from the contracting firm and certain guidelines had to be followed with respect to the nature of production. This buyer would frequently request that the firm manufactured variations in the sweets which it made, usually changes in the flavourings, shapes, colours and names, yet such directives were almost always accompanied by detailed instructions as to the manner which this should take and the ingredients which were permissible in the alteration. Therefore despite the introduction of variations in product lines there was a distinct absence of R&D activity, this being undertaken by the contracting firms and subsequently the rights to the new products were exclusive to that company.

The growth of the company was also limited in that, all the produce manufactured for the contracting firm was under the larger firm's trade mark, and therefore this prevented the economies which could have arisen through advertising at the point of sale with the consumer. The firm was virtually an appendage of the contractor with lower order functions such as manufacturing occurring here, while the higher order functions such as R&D and marketing occurred within the contracting firm.

Although employment within the firm in 1988 was 30 employees and by 1993 it was 37 employees, the composition of employees had changed substantially over this period with a significant rise in the proportion of part-time employees. There was a high labour-turnover within the firm, with approximately 7 employees having left the firm during the 12 months prior to the interview.

Additionally, as all the jobs for which part-time labour was recruited were low-skilled, this facilitated the high turnover of staff throughout the production line.

Case Study Five:
A Northern Ireland firm within the electrical and electronics industry was established in 1974 and by 1988 had an annual turnover of £2.2 million. Turnover increased by 13 per cent in real terms between 1988 and 1993 whilst employment increased by 6.8 per cent. The firm distributed 10 per cent of its output within Northern Ireland, 70 per cent within Great Britain, and an additional 15 per cent was sold outside the British Isles, namely to France and Germany. Subcontracting relationships with three firms represented 50 per cent of total sales, a market position which the firm believed provided them with relative stability.

Little evidence was found of numerical flexibility within the firm, with all staff being employed full time and on permanent contracts. Therefore the extent to which the firm was pursuing the minimization of costs through the restructuring of the workforce, was limited.

Although the attainment of 'growth' was not the primary objective of the firm, the maintenance of existing products in existing markets was an overriding concern, with the belief that it was only through the achievement of BS 5750 that the contracting relationships in particular, could be protected. As these orders accounted for approximately 50 per cent of the gross turnover of the firm, the realization of this objective was paramount. Yet the management expressed the view that while BS 5750 was of central importance for the sales to the contracting firms, such was not the case for the majority of the other sales. Instead, characteristics such as the price of the boards and the fulfilment of predetermined delivery times, along with the overall quality of the products, were the determining factors in the renewal of orders. Yet the management highlighted that in the long-term, quality certification would be equally as important for the 'other' buyers, as for the contracting firms.

Throughout the interviews, a number of firms in Northern Ireland expressed their contempt for the quality standard of BS5750 and were reluctant to engage in the formal attainment of certification. However this firm although conscious of the financial costs and administrative adjustments which would have to be made, in the attainment of BS5750, also acknowledged the competitive advantage associated with obtaining the quality standard before their competitors. Indeed, the firm intended strategically to use the quality certification as a lever in expanding their market, both within the local area and beyond.

Although the firm was not actively engaged in employee training, it perceived the need to train to be cumbersome and financially draining on the firm, especially as no short-term gains were evident from training. Such a view is however erroneous as was previously outlined and indeed it would appear that the desire for BS5750 had virtually 'blinkered' the firm to the alternative forms of potential growth.

In terms of Just-In-Time and the elimination of stocks, it was apparent that the firm was attempting to increase the responsiveness of production to changing demands for various types of circuit boards. Yet this had not eliminated the maintenance of stocks of basic circuit boards within the firm. In terms of a system of total quality, the management intended to reduce the reject rate of production, and eliminate the costs associated with the wastage of materials. The management stated that this higher level of observation and awareness by shop-floor employees - along with the eventual introduction of advanced quality checking machinery - would significantly reduce the reject rate at the end of the production process from its present level of 3 per cent.

Where a proactive strategy was adopted in counteracting a weak market position such as through the attainment of quality certification (Case Study Five) this was seen to result in higher growth, as measured by turnover and output per employee from 1988 to 1993, than as manifest through the protectionist strategy (Case Study Four). These findings suggest that where a firm either remains within the quagmire of past operational policies or adopts a strategy which shelters it from strategic change in the short term, then this will culminate in negative growth. Small firms must therefore seek ways by which they can proactively improve their competitiveness and thus increase their relative market share.

Numerically flexible work practices

Numerical flexibility is identified as the ability of firms to change their labour inputs in response to changes in the market. In this way, labour behaves like any other commodity within the firm by being employed and deployed when needed. The incorporation of numerically flexible work practices within a company can be seen to provide a competitive advantage These advantages are most evident in terms of enabling management to establish a point where X-efficiency is at its lowest through both the minimization of labour costs and the maximum efficiency of the labour to capital ratio.

Belous (1989) highlights that many companies now use part-time employees as an alternative to boosting labour costs in light of shifts in labour demands and / or labour shortages. In addition, he notes that within the USA, temporary employment is the fastest growing element of the contingent workforce. While there are a number of reasons for the increasing incidence of 'contingent labour' (most notably the establishment of national labour markets which are conducive to flexible work practices - through the devolution of the trade unions and increasing unemployment rates) it is the increasing competitive pressures within the firm which have resulted in companies continually seeking ways to minimize their costs. Through this manufacturing efficiency can be improved and profit margins increased.

While a firm may gain cost and subsequently competitive advantages from the implementation of numerical flexibility, the firms ability to introduce these labour practices may be constrained by the industrial sector and the market environment within which the firm is operating. For example, the sample results pertaining to the food processing sector demonstrated that there was a similar proportion of part-time employees within both Northern Ireland and Massachusetts, equating to 12.5 per cent of the workforce with these workers being concentrated within the low-skilled sector of the firm. Conversely within the higher skilled sectors it was evident that where temporary or part-time workers were being employed this represented a smaller proportion of the overall workforce. Temporary and part-time employees are therefore associated with low-skilled jobs and therefore the ability of a firm to introduce such workers will be a function of the incidence of low-skilled jobs within the firm.

Firms which require a high level of skills from all of their workers may however incorporate numerical flexibility if there is either a highly skilled and accessible workforce external to the firm or if the firm complements the use of temporary and part-time employees with a comprehensive training system. Indeed results from the matched plant survey illustrate that it was only through the introduction of comprehensive forms of training and innovative and progressive methods of production (as manifest through Just-In-Time principles and the formation of Quality Control systems) in conjunction with numerical flexibility that sales per employee increased and thus profit margins as a proportion of labour costs were improved (Case Studies 6, 7 and 8).

Case Study Six:
A Massachusetts firm established in 1982 manufactured pipe welding and process piping. Although employment had risen to 30 full-time employees by 1988, this had declined to 25 by 1993 with turnover increasing by 45.4 per cent over the same period.

Forty per cent of the manufactured goods remained within the local state boundaries of Massachusetts, with 20 per cent going to other states in New England, and 40 per cent to other areas throughout the USA beyond New England. The wide distribution of sales was attributed by the management to their competitive advantage in terms of quality of workmanship. Therefore it is plausible to state that the quality of the products was a key component in assuring their marketability both locally and regionally.

With regard to the metal products sector, the Massachusetts firms displayed a lower level of dependency upon subcontracting relationships than the Northern Ireland firms. For example, this firm had established subcontracting contracts with five firms and this represented 50 per cent of their sales output.

Conversely, as was found for some of the Northern Ireland firms, dependency upon subcontracting relationships represented 80 per cent of total sales output.

With respect to the employment structure of the firm it was found that the reduction in the number of employees was characterized by a change in the nature of employment and the, somewhat tentative, adoption of numerical flexibility. The proportion of people employed within different sections of the firm had remained constant from 1982 to 1988, however since then the firm had introduced temporary contracts of employment. Of the five temporary employees within the firm, two were shopfloor machine tool operatives and the remaining three were assemblers and fitters, with each having employment contracts for three months at a time. The management indicated that during the year prior to the study, as a result of a high level of sales growth none of the five temporary workers had experienced a period of unemployment. Yet, despite the permanency of these jobs during the previous 12 months they found that it was essential to maintain temporary contracts as it cushioned the firm against fluctuations within the market.

It is the combination of the introduction of temporary contracts of production - which may be terminologically identified as the incorporation of numerically flexible forms of production at the micro-level - along with the multi skilling of employees which helps to explain the change in turnover relative to employment[4]. It was found that the firm spent $14,500 on the training of employees in the 12 months prior to the interview, and the management stated that this figure was actually lower than the previous two years due to the repayment of a loan which was obtained to expand their premises. Of the training undertaken in the previous year only $2000 had been spent on the training of management with shop-floor workers and fitters receiving the majority of the training budget. Of the 13 shop-floor workers, 2 of these employees were being partly financed by the firm to undertake a degree course which would upgrade their qualifications from diploma level. As a consequence of the large investment by the company which this was entailing the firm had drawn up contracts with these employees which ensured their continual employment with the firm for a period of five years on completion of their degrees.

It can therefore be seen that the development of a skilled 'core' group of workers was being established within the firm. Yet it is important to note that these 'core' workers were not being trained at the expense of more job specific and 'peripheral' workers, such as the temporary employees, but rather these employees were also receiving training. The management also highlighted that they had the objective of promoting the flexibility of workers throughout the firm, with even secretarial staff being encouraged to spend one working day every six months on the shop floor. Thus, a system akin to that of functional flexibility was aspired to within the firm, yet this was not confined to the core workers, but rather was a characteristic feature of all employees. It is therefore possible to suggest that the rising turnover level within the firm bore a positive correlation to the restructuring of the labour force over the 5 years, from 1988 to 1993.

In this context it is also important to note that the firm had only 10 per cent of its total capital stock under 5 years of age. It did however have a number of numerically controlled machine tools in operation within the firm - most of which a large proportion of the employees were trained to use. The belief of the firm that the quality of its goods represented a key competitive advantage over its rivals was supported by the operation of groups of employees working together on various stages of production and thus monitoring the standard of output in a manner akin to that of Quality Control Circles.

In summary, this company characterized the predominant features of the Massachusetts firms within the industrial sector of metal goods. It was apparent that the introduction of a functionally flexible system was a key element in the development of the firm especially since 1988. While it is sometimes expected that where functional flexibility is introduced, this will be complemented by an increase in the utilization of part-time labour, this was not found within the firm under investigation.

This can perhaps be best explained by the nature of the industry, in that increasing complex production methods demand that employees are fully trained and skilled in the aspect of production which they are employed in. Further, the type of job orders which would be attained by the firm would have deadlines with respect to the delivery or installation of products and therefore it would be essential that employees could be hired whenever necessary to complete orders on a full-time basis and laid-off during periods of diminished demand. Therefore temporary contracts are the most cost efficient form of production for the manufacturing firm.

Findings from the matched plant survey demonstrate that there are sectoral disparities in the incorporation of numerically flexible labour practices with a bias being evident in favour of industries with a significant concentration upon low skills, i.e. food processing. Conversely a regional bias was evident within the engineering sectors whereby it was predominantly the Massachusetts firms which were introducing temporary and part-time contracts. It is significant that the Massachusetts engineering firms had a lower employment to turnover growth ratio over the period and a higher growth in sales per employee for the engineering sector as a whole (Table 4.3). While numerical flexibility may have been important in reducing costs it does not account for the higher relative growth in turnover amongst the Massachusetts firms over the period. What were the other mechanisms at work within the firms which were contributing to higher turnover growth levels within the Massachusetts firms? The next two sections briefly address the issues of training and the resultant functional flexibility within the workforce and the incorporation of innovative and flexible principles of production.

Worker Training And Functional Flexibility

While Functional flexibility[5] and worker training are often regarded as two distinct elements it is unequivocal that the two are intricately linked. Within small firms by virtue of their size there is frequently an overlap of functions, however this is generally more applicable to the management as opposed to the shop-floor workers.

Although it is beneficial for a small firm to have a flexible and adaptable management team, the principles of functional flexibility extend throughout the work place. Shop floor employees must display efficiency through the production process and indeed attain an understanding of the underlying mechanisms and strategy at work within the firm. Where such a position is achieved then systems can be set in place which facilitate the transferability of both workers and knowledge through the firm in a manner which identifies and counteracts inefficiencies. Therefore as with the introduction of numerical flexibility, a position can be approached which enables X-efficiency to be eradicated and productivity gains made.

As a result of the ambiguity of the term 'functional flexibility' it was difficult to isolate its presence within the firms in terms of a specific quantifiable measure. However, results from the sample demonstrate that where functional flexibility was introduced as manifest in the existence of work teams, interdependence between workers and low levels of demarcation of tasks etc. a substantial investment in employee training was also found (Case Study Seven).

Case Study Seven:

A Massachusetts firm within the sample was established in 1960 in the heart of Massachusetts, namely Cambridge, Boston, and manufactured elevators, dumbwaiters and stair climbers. In 1988 the total sales value of the firm was $3,000,000 with this increasing to $4,000,000 by 1993. Allowing for inflation[6] this meant that turnover had increased by 20.29 per cent over the period. With respect to employment, the numbers employed by the firm had remained constant at 35 employees between 1988 and 1993, yet a change in the composition of the work-force had occurred with the introduction of 2 part-time workers - equating to the reduction of one employee in terms of man hours. Additionally the total number of employees directly involved in manufacturing upon the shop-floor had declined with the introduction of 3 employees dealing exclusively with customer service, marketing and sales.

The management stated that the attainment of growth was not the primary objective of the firm, and instead placed greater emphasis upon the reduction of costs. This firm was changing its remit in two ways, firstly through the adoption of a more *aggressive* marketing policy and the employment of specialist staff in this field, and secondly through a more *competitive* framework of production, by the minimization of costs, both in terms of capital and labour requirements.

It is perhaps with regard to the amount of money which had been invested by the firm in the training of employees during the 12 months prior to the interview, which is demonstrative of the desire of the firm to improve its competitive position. In total, $34,500 had been spent by the firm on the training of fitters, electricians, sales staff and management. This training was external to the firm and consisted of fulfilling the licensing requirements for electricians within the state of MA; providing sales seminars for the sales and marketing staff; and a three week course on the technological advancements within the elevator industry, for one of the managers. Additionally all staff received on-the-job training from managers and frequent assessment of work ensured the capability levels of all employees. Perhaps as a consequence of the high skill ability of the workers, there was a strong emphasis on sharing tasks and team working. The manager with whom the interview took place, stated that the interdependence of workers had two important effects, namely the monitoring of the quality of production and the monitoring of time management throughout the work force. Therefore the introduction of functional flexibility and thus the enabling of a more flexible form of production, had been introduced to encourage a more efficient use of both capital and labour. Further in acknowledging the crucial importance to the firm of multi-skilling the work force the manager also propounded the financial burden which training imposed upon the financial resources of the firm. Yet this financial burden was seen as a short-term and necessary evil which would culminate in a beneficial and lucrative advancement of the firm in the long-run.

With regard to the capital stock, approximately 50 per cent was over 10 years of age with only 15 per cent of new capital having been acquired by the firm in the previous five years. Of this capital stock only a small proportion consisted of technologically advanced machinery, such as CNC machine tools. While the firm attempted to formulate its production methods around the principles of JIT the interviewee stated that a number of refinements of the production process were presently under consideration - one of which was the costs associated with the attainment of components parts from suppliers. Further, although the firm did not participate in research and development, it felt that this was a very important aspect in the long-term future of the business. Financial constraints - identified by the managers as repayment on loans for capital equipment, along with the investment in training, and a small profit margin - were highlighted as the predominant reason for the inability of the firm to undertake either process or product R&D.

Where there was evidence of management attempting to introduce functional flexibility amongst their workers then training was generally perceived as an integral aspect of the strategic development policy. It was noticeable however, that while a 'core' workforce was evident in the disproportionate investment by some of the firms in their managerial and high skilled employees, within other firms the training policy explicitly incorporated the un-skilled and semi-skilled employees (Case Study Eight).

While there was comparability within the food processing sector between the two regions in the incidence of numerical flexibility, such was not as apparent with regards the functional flexibility of the workers. Within the Northern Ireland food processing firms there was a clear demarcation of tasks, with the preparation, processing and presentation aspects of production being clearly demarcated with low levels of transferability between these sections. Alternatively within the Massachusetts firms it was frequently found that during the initial period of employment for new-recruits - which usually consisted of 6 weeks - a component of their induction training was to gain an appreciation of all aspects of production, from the receiving of raw materials to the point at which the finished product left the factory. Further, the workers within the Massachusetts food processing firms were expected to clean and maintain the machinery on which they worked, and should any irregularities in production become apparent, notify the foremen or managers, so that problems could be quickly identified and steps taken to reduce both the down-time of production associated with breakdowns, and the wastage of resources, through the minimization of inferior products.

Conversely within the Northern Ireland firms, maintenance work was viewed as the function of engineers and technicians with problems in the equipment being perceived as the sole responsibility of these workers. It is interesting to note that in aggregate the Massachusetts sector had fewer machinery breakdowns during the three months prior to the interviews than the Northern Ireland firms, and this can be assumed to be a direct function of the maintenance of the capital stock by shop-floor employees, as the differences between the regions in the age of the machinery were negligible (Chapter Seven).

While sectoral differences were more important than regional differences in influencing the implementation of numerically flexible employment practices, regional differences were more apparent in the implementation of functional flexibility. Expenditure on training per employee was considerably higher within the Massachusetts firms and this would suggest a direct relationship between training and functional flexibility. It is perhaps the differences between the food processing firms within Northern Ireland and Massachusetts along with those firms within the electrical and electronics sector (Case Study Five and Case Study Six) which demonstrate most clearly the additionality to turnover growth of higher expenditure on employee training and the encouragement of transferability throughout the work place.

Innovation, just-in-time and total quality production principles

It is a commonly held view that small firms are important innovators within an economy the origins of such a belief often being attributed to Schumpeter (1934) who described them as being capable of initiating gales of creative destruction. It is unquestionable that if small firms can develop new products and thus establish competitive advantages - at least in the short-term - then they will experience growth at a faster rate than the non-innovators.

This competitive advantage is not confined to product innovations but also encompasses process innovations whereby systems may be established within a firm which facilitate a more efficient utilization of resources (Drucker, 1985). Therefore innovation and flexible production systems as characterized by Just-In-Time and Total Quality, are not mutually exclusive, in that, while innovation may refer solely to products it also includes changes within the production process. Where productive inefficiencies can be identified and steps taken to counteract these then this will result in the elimination of X-efficiency and the attainment of higher profit margins which through re-investment facilitate growth in the long-run.

Results from the matched plant sample demonstrate that where innovative and flexible principles of production were present within the firms, this was associated with a higher level of turnover growth from 1988 to 1993 (Case Study Eight).

Case Study Eight:
The firm under investigation within Massachusetts, was established in 1982, yet by 1988 had an exceptionally high level of turnover, of $20 million. By 1993 this turnover had increased to $35 million, which equated to a real increase in turnover of 70.6 per cent. The products which the firm manufactured consisted of semiconductors, micro-processors, resistor networks and capacitors. It should be noted that a number of these goods are high-value goods, and therefore, in part, this accounts for the phenomenally high level of turnover relative to the majority of firms within the sample population. It is also interesting to note that the key aim of the firm as stated by the management was to stimulate the development of new products. This in line with recent research on product innovation (Roper et al., 1996) assists in the understanding of how the firm has grown so quickly, not only from formation, but also since 1988.

Numerically flexibility is a crucial aspect of the production of the firm especially with respect to the presence of subcontracting relationships and the strategic alliances which these incorporate. For example, 90 per cent of total sales was attributed to 6 companies with which the firm has a contractual agreement. These relationships were strategic in the implications which they held for the undertaking of R&D. The firm had an established policy of undertaking R&D alongside two of the firms for which it contracted production, and this frequently manifested itself in the ability of the 'small' firm to use the R&D facilities of the larger companies as well as obtaining access to their specialist personnel during the early stages of product development. The only conditions associated with this for the small firm, were that the host company had the authority to share in the patented rights of any developments in either the hardware or software developed by the firm. The management of the firm highlighted that although they were moderately content with the prevailing conditions associated with the undertaking of R&D, they would prefer to have independence from the legal and contractual rights of these larger companies, yet this was impossible due to financial constraints. However, the independent execution of R&D was perceived as a viable long-term objective, when profit levels were high enough to merit investment in the requisite capital equipment and skills.

In terms of employment, there was a positive percentage change of only 14 per cent over the five year period, equating to a rise from 28 employees in 1988 to 32 in 1993. Further, it was demonstrated that the firm had introduced part-time workers during this period consisting mostly of un- and semi-skilled employees, who had a standard working week of 23 hours. It was noted however that these part-time workers frequently worked full-time shifts, with two out of the three weeks prior to the interview being characteristic of this. The firm did not employ temporary workers and had displayed no intention to introduce them into the employment structure of the firm.

With respect to the skill levels, training and functional flexibility of the employees, the presence of a highly skilled workforce was apparent. Six semi-skilled workers had completed certificate courses in electrical engineering and associated subjects, while 15 skilled shop-floor employees had obtained diplomas and taken part in advanced courses of study. There were four fully qualified and licensed electrical engineers within the firm, with all three managers also having degrees in electrical engineering. In addition, one of the managers had obtained an MBA and another had a post-graduate Master's degree in computers and business administration. Further, there was a licensed accountant within the firm (who regularly assisted on the shop-floor doing low-skill tasks), and three sales and marketing staff, who were all qualified to degree level.

During the 12 months prior to the interview a total of $28,200 had been spent on the training of employees, covering such areas as the introduction of new technology; the fulfilling of licensing requirements; marketing courses; and specialist conferences in electrical engineering etc. While the majority of formal, off-house training was undertaken by skilled, technical and managerial employees, the firm also encouraged the training of un- and semi-skilled employees in-house, both through on-the-job training by engineers and short instructional videos within the 'training room' of the firm. It is of interest to note that of all the firms visited, this was the only firm which had a 'training room' and formally undertook to train the unskilled employees within the firm.

The management stated that there was considerable overlap of functions within the firm, with skilled employees being adaptable throughout the work-place. Electrical engineers were seen to work in teams ranging from two to three employees to much larger groups of seven to eight. Within these groups a multiplicity of functions was performed, with briefings being a prominent characteristic of the firm, whereby all the engineers would gather for between 15 and 30 minutes on three mornings each week to discuss production, technical problems, developments and areas of improvement. The interviewee stated that sometimes the necessity for these meetings was minimal, but they still occupied a central position in the day-to-day life of the firm, as they were perceived as providing an essential form of communication between different sections of the workforce, while stimulating enthusiasm and innovation, in that every part of the firm could see that it had an important role to play in ensuring the efficient and effectual functioning of the whole firm.

As part of the team-work structure within the firm, the formulation of a policy of TQM had evolved. The management stated that this had been a fundamental aspect of production since the inception of the firm, and consequently it displayed a reject rate at the end of production of less than 0.5 per cent. The firm did not have a policy of manufacturing for stock, but instead indicated that it was capable of quickly adapting production in-line with changing market demands. Therefore the firm was reducing costs through the elimination of buffer stocks. Yet what was of particular interest was the declaration by the management that they perceived their growth and prosperity within the electronics market to be directly related to their position within Massachusetts. They believed that their local access to a plethora of advanced electrical and electronic, and high-technology firms within Massachusetts, and especially along Route 128 to be a competitive advantage which had assisted in the establishment of strategic alliances with these larger companies and had facilitated the setting-up of R&D activities which would otherwise have been impossible due to the high capital investment which such demands. It could have been expected that where there was a high level of competition with a number of firms manufacturing similar products that this would have saturated the market, yet conversely the opposite was found with increased competition generating an environment of co-operation and innovation, with such being translated into a culture of assertiveness and aggression within the firms.

Therefore within this firm there was a high level of growth in terms of turnover, however this was not translated into comparable changes in employment. If aspects were to be accentuated in an attempt to explain the high rate of growth within the electrical and electronic sector, as epitomized by this Massachsuetts firm, then a number of factors could be identified.

Firstly, perhaps the most important factor which contributed to the growth of the firm was intangible and related to the competitive and aggressive spirit of the workforce. Within the firm there was an explicitly charged atmosphere of advancement, and such was most apparent through the aspirations of the firm to develop new products.

Secondly, there was close co-operation with larger companies which facilitated the establishment of strategic alliances whereby the resources of both the large and the small firms could be pooled to advance the development of products.

Thirdly, the firm had identified the benefits which could be realized from the restructuring of the workforce and therefore had introduced numerical flexibility in the form of part-time workers and functional flexibility in the multiplicity of tasks which a range of people could perform.

Fourthly, there were distinct aspects of JIT through the virtual elimination of buffer stocks and the relative geographical proximity of buyers thus enabling the firm to minimize the turnover time of production, that is, the length of time it took from the initial manufacturing of the product to the point of delivery with the customer was reduced to the bare minimum.

Finally, the firm stimulated innovation and development through the utilization of team work and QCC's thus enabling the effective operation of a system of TQM, which attenuated the reject rate of production and thus minimized the wastage of resources.

Innovation whether product or process related undoubtedly encourages efficiencies which culminate in profit growth. Where a firm seeks to improve its responsiveness to market changes and reduce its costs through the elimination of wastage and stock then this will undoubtedly lead to the attainment of a more competitive position.

Summary

In conclusion it is clear that there are a range of mechanisms within the production process which have diverse and profound implications upon the changes in turnover and employment. With respect to the initial proposition as postulated by North and Smallbone (1993:4) that there was a direct and positive correlation between output and employment, that is, that, 'job creation was closely tied to the growth of the firms in output terms (measured in terms of deflated sales turnover 1979-90) and that nearly all the new jobs were created by firms which at least doubled their output over the decade', it has been demonstrated that within the context of this analysis, these results are not necessarily correct. Instead, the findings of this research demonstrate that there is not a direct relationship between turnover and employment as suggested by

North and Smallbone (1993) but rather there are a plethora of factors at work which act to complicate the relationship.

This does not however eradicate the fact that in general a firm will only increase the size of its workforce when it experiences an increase in output as measured by the volume of sales, as opposed to the value. Yet, there are examples as demonstrated, when the nature of employment has increased through the incorporation of a policy of numerical flexibility and therefore while in absolute worker-hours employment may have only increased by one or two full-time employees, this may be manifest in the absolute employment of four to five employees. Hence, where this has occurred this will produce a disproportionate increase in employment relative to turnover and the presence of anomalies within the data may result. Additionally, there were cases where total turnover values decreased while employment levels increased. In some of these cases the product range of the firms and the nature of production had altered substantially between 1988 and 1993, with the introduction of service activities being apparent within a number of Northern Ireland engineering firms. In such circumstances employment increased to facilitate such changes, while the returns to these changes had not become evident.

In addressing the four questions which were posed at the beginning of this chapter the findings can be summarized as follows. Firstly, there was significant evidence of growth within the matched plants. Secondly, regional disparities were manifest in the extent of growth as measured by turnover and employment, with the Massachusetts firms generally displaying higher levels of turnover growth than their Northern Ireland counterparts. However, the rates of turnover growth and employment growth did not display any direct relationships between the regions and this would therefore lead to questions regarding the legitimacy of regional development founded upon the principles of employment generation.

Thirdly, there were disparities in the extent of growth between the sectors of analysis. Growth, as measured by changes in employment, was greater in the food processing sector than in engineering for both regions. However, in terms of turnover growth the food processing firms grew faster than the engineering firms in Massachusetts, while the reverse was found within Northern Ireland. It should be noted that there were also further sectoral disparities within the engineering firms. For example, although the food processing sector displayed higher rates of growth than the engineering firms in Massachusetts, those firms within the electronics sector increased their real turnover by a greater margin than either the food processing firms or the remaining engineering firms. Again the non-complementarity between turnover and employment growth has been emphasized as in the case of the Northern Ireland sample whereby employment has increased disproportionately to engineering, but this was not matched by a similar growth in turnover.

The final question which was posed at the beginning of this chapter queries the reasons for growth disparities. From the case studies it would appear that the attainment of growth within a company is multi-faceted and complex however generalized patterns were apparent. For example, the restructuring of product and service markets led to a decline in turnover, however, it is possible that in the long-term turnover would increase as firms strengthen their position within the new market. Firms which failed to make any changes to their internal strategic or organizational environments also were seen to decline over the period of analysis, and this accentuated the dynamic nature of competitiveness. In contrast, where a pro-active strategy was adopted to counteract market vulnerability growth as measured by turnover and output per employee increased significantly.

Numerical flexibility was seen to be present within some firms more than others, however, it was concluded that while numerical flexibility may have been important in reducing costs, alone, it did not account for the higher levels of turnover growth. However, important disparities were present between the two regions in the adoption of functionally flexible work practices and it was demonstrated that there was an obvious link between this and turnover growth. Finally, it was also found that firms which were innovative and had adopted production strategies based upon the principles of flexibility experienced higher levels of growth than non-flexible firms.

In summary, examination of growth levels within the matched plants has established that there is not a direct and positive correlation between turnover and employment, and indeed a significant negative correlation was identified. Additionally, while the short case studies have attested to some of the characteristics underlying changes in turnover and employment growth[7], it is only by a thorough examination of the manifestation of flexibility within the labour and production process that the existence of a relationship between relative performance and flexibility can be identified. Chapters Five and Six analyse the extent of labour flexibility within the sample firms while Chapter Seven examines flexibility within the production process.

Notes

1 Other variables such as age of the firms, employment practices - ie. utilization of part-time, temporary labour etc. - job demarcation and interaction of workers, extent of employee training, skill levels of employees, methods of production and the impact of external agencies such as 'Quasi-Autonomous Non-Governmental Bodies' and other official agencies are also examined to provide a comprehensive understanding of the factors determining the extent of growth within the firms.

2 Nineteen-ninety-three turnover values were converted into 1988 prices by adjusting for changes in the Producer Price Index. Producer price indices were used in accordance with

the industrial sectors under investigation, and a weighted average was calculated. The percentage change in the Producer Price Index from 1988 to 1993 was as follows:

Metal Products, Northern Ireland	+22.9 per cent;
Metal Products, Massachusetts	+3.13 per cent;
Mechanical Engineering, Northern Ireland	+27.70 per cent;
Mechanical Engineering, Massachusetts	+10.84 per cent;
Electronics, Northern Ireland	+18.00 per cent;
Electronics, Massachusetts	+2.53 per cent;
Food Processing, Northern Ireland	+24.07 per cent;
Food Processing, Massachusetts	+6.38 per cent.

3 The extent to which the SIC's have experienced growth over the five years prior to the study is outlined, however the determination of growth in this instance is solely through the analysis of employment data.

4 That is, a restructuring of the firm occurred during the five year period from 1988 to 1993, as manifest through the reduction in total employment and the incorporation of temporary contracts of employment. Additionally the enskilling of the remaining workers facilitated efficiencies within the firm which had hitherto been unrealised.

5 Functional Flexibility is defined in terms of the ability of firms to adjust and deploy their employees' skills in line with changing production methods and market demands.

6 That is the PPI at a rate of 10.84 per cent.

7 Growth in this instance being measured by the real change in turnover between 1988 and 1993.

5 Functional flexibility in the production process

Introduction

The regulation school of thought identifies five main areas wherein flexibility may be deployed (Moulaert and Swyngedovan, 1989), these include; the production process, the labour process, space, the state and the over-riding ideology. For those flexible specializationists however, it is flexibility within the sphere of production which is fundamental to overcoming the rigidities of the Fordist era. Implementing flexibility in production embraces the underlying principles of production in terms of the utilization of fixed capital, and focuses special attention on the organization of work and the skills of employees.

Central to the debate on flexibility in the production process is the traditional labour market segmentation theory as outlined by Doeringer and Piore (1971:165). This recognized a primary labour market which corresponded to the 'Fordist' sector and a secondary labour market which corresponded to the competitive sector. The primary sector consisted primarily of production workers and an 'independent class' consisting largely of managers and professionals. The secondary sector was characterized by 'low wages and fringe benefits, poor working conditions, high labour turnover, little chance of advancement, and often arbitrary and capricious supervision' (Doeringer and Piore, 1971:165).

More recent models have proposed alternative decompositions of firms' work forces. For example, Atkinson (1985) suggests, that in the 1970s many firms began to pursue a policy of flexibility resulting in 'horizontal segmentation into a core work force which ... conducts the organization's key firm-specific activities surrounded by a cluster of peripheral groups. Their twin purpose is to protect the core group from numerical employment fluctuations while conducting the host of non-specific and subsidiary activities which all organizations require and generate' (Atkinson, 1985:14). Atkinson argues that because firms have put a premium on achieving a workforce which can respond quickly, easily and cheaply to unforeseen changes a new form of labour-market segmentation has arisen between a functionally flexible core labour force, and a numerically flexible peripheral one. This has led to his identification of three kinds of flexibility, namely numerical flexibility, functional flexibility and financial flexibility. The framework developed by Atkinson (1985), and subsequently applied empirically by Atkinson and Meager (NEDO, 1986) has become the benchmark against which labour flexibility is gauged.

Atkinson's 'micro dual labour market model' (Pollert, 1988:283) of the flexible firm proposes that the core workers will be functionally flexible while the peripheral workers will be numerically flexible. The core workers are central to the firm's operations and are differentiated from the peripheral workers in terms of pay, status and job security. Better job security amongst core workers is due to their higher skill levels and the limited availability of similar workers in the external labour market.

As functional flexibility is a particularly nebulous concept problems have emerged in identifying and quantifying its presence. Conversely, the ease by which measures of numerical flexibility can be obtained, has often led to the evidence of numerical flexibility being over-emphasized relative to functional flexibility.

In acknowledging the symbiosis of numerical and functional flexibility within the flexible firm model analysis is made of both forms of labour flexibility within the matched plants. This chapter focuses upon the presence of functional flexibility within the matched plants, while chapter six examines the presence of numerical flexibility within the firms.

In determining the ability of the firms to introduce functional flexibility, the aptitudes of the employees within the Northern Ireland and Massachusetts firms are examined in terms of qualification levels and investments in training. Each of the skill groupings are then examined separately. The extent to which employees were functionally flexible is determined, through the undertaking of multi-tasks; the elimination of job demarcations and the existence of continuous on-the-job training. The relative importance of supervisory workers within the management hierarchy for both regions and sectors is then outlined, in conjunction with the perceived role of supervisors within the firms; the extent of training; the nature of training; the skills acquired through training and the adaptability of supervisors throughout the workplace. The final part of the chapter focuses upon the managers within the firms. Again, the qualification levels of managers are examined in conjunction with the investments in managerial training; the nature of this training; the skills acquired through training and the flexibility of managers throughout the workplace.

Both this chapter and chapter six consolidate and augment the work of Atkinson and Meager in establishing whether or not numerical and/or functional flexibility was evident within the sample of small engineering and food processing firms in Northern Ireland and Massachusetts. Whereas Atkinson focused upon large firms and took for granted the inherent flexibility amongst smaller firms, this investigation in focusing upon the small firm sector seeks to test the applicability of the labour flexibility thesis to small firms. Additionally, by focusing upon the same industrial sectors as Atkinson, namely engineering and food and drink, this controls for variability in the adoption of flexibility across industrial sectors.

Functional flexibility

A working definition

Functional flexibility can be defined as, 'The ability of firms to adjust and deploy the skills of their employees to match the changing tasks required by a changing workload, production methods or technology' (Pinch, Mason and Witt, 1991:208). While numerical flexibility enables firms to respond to changing market demands, functional flexibility is more closely linked with changes in the operational environment of the firm. Economies of scope, advanced technology and changes in the organization of production interact to necessitate the flexible deployment of skills.

While the definition of functional flexibility refers to the ability of firms to deploy employee skills to changing tasks, this ability is dependent upon the characteristics of the workforce. For example, if workers are low-skilled and unable to adapt to changing tasks then firms will be constrained in their ability to introduce strategies aimed at improving efficiency. Conversely, it is plausible that employees may be highly skilled and capable of being functionally flexible, yet multi-skilling and the crossing of occupational boundaries are not encouraged as part of the production process. It is prudent therefore to focus initially, upon the potential ability of firms to deploy their labour in a flexible manner, as manifest through employee skills.

The AEU have linked functional flexibility to 'dual skilling, cross skilling, multi-skilling, craft/production interchange etc.' (AEU, 1992:4). The AEU's policy for controlling the introduction of a flexible workforce within a firm, highlights that almost all workers have a skill at which they are competent, and through the 'adding-on' of skills through 'appropriate training and consolidation' (AEU, 1992:2), workers can become more 'versatile and valuable to the employer' (ibid). Of central importance in the attainment of greater functional flexibility, is the training of the workforce to a point where they are 'multi-skilled'. Although the centrality of training in functional flexibility is explicit, the definition ascribed to functional flexibility can result in training being overlooked. Changing demand and production forces within firms and the introduction of new technology provides the impetus for creating a core workforce which is functionally flexible. Yet, functional flexibility can only be attained through the training/re-training and multi-skilling of employees.

Through the implementation of multi-skilling, workers will be able to respond to any 'changes in the established working practices which would result in technicians, craftsmen or production operatives taking over tasks previously carried out by specific individuals or groups as a job core activity, or vice versa' (AEU, 1992:4). Atkinson and Meager (1985) argue that, in the attainment of functional flexibility, it is only the 'core' workers who are multi-skilled, and instead of all workers increasing their skill capacity, the peripheral workers are further marginalized within the company as they receive little or no training to improve their competence base. Conversely, the AEU see multi-skilling extending to all employees irrespective of initial competence levels and instead of the development of a core-periphery workforce, there would be an upgrading of skills for all employees. Hence, through a multi-skilled workforce (including all employees) workers would become more efficient, effective and productive as evident in higher levels of added value, higher profit per employee and fewer inefficiencies within production (see Chapter Seven).

Beyond the direct implications of training on functional flexibility as manifest within the flexibility literature other work presents an implicit assumption concerning the direct correlation between training and competitiveness. For example, the Office of Technology Assessment (1990:1) states that; 'Training goes hand-in-hand with productivity, quality, flexibility and automation in the best performing firms...Good training pays off - for the individual worker whose skills are upgraded, for the company seeking a competitive edge, and for the nation in overall productivity and competitiveness'. Within this statement training is seen as the 'key' to rejuvenation, not only on the micro-level of the firm, but also in terms of national competitiveness.

This direct link between productivity and training, and its effect upon the competitive performance of national economies is emphasized by the National Institute for Economic and Social Research. They state that; 'International productivity differences can arise for many reasons and an exhaustive explanation may never be possible. Nevertheless, detailed studies have shown that training is the key to higher productivity in many industries and occupations. It is now recognized as a precondition for the attainment of higher living standards...' (National Institute Economic Review, 1991:7).

A further study by the National Institute for Economic and Social Research (Prais and Wagner, 1988), through a matched plant analysis, compared the relative performance of firms within the UK, France, Netherlands and Germany. From examining physical output and the vocational qualifications of employees they demonstrated the effect of various skill levels upon quality and productivity. They found that within the clothing industry, the better qualified first-line workers within the German firms were able to overcome both complex problems and processes while simultaneously maintaining a consistently higher standard of both productivity and quality of products relative to the matched plants in the UK, France and the Netherlands. Additionally, the German workers could undertake a broader range of tasks than their British counterparts, which included routine and preventative maintenance of capital equipment. In general, the work forces within the German plants were better equipped to respond to technological and organizational changes within the firms with supervisory workers being better qualified than their international counterparts.

Within the context of the Republic of Ireland, the White Paper on Manpower Policy (p.1 4306, September 1986, Para. 4.1 - as quoted in Roche and Tansey, 1992:41) stated that: 'The success of the Irish economy depends largely on the success of our enterprises and that, in turn, is crucially dependent on the knowledge, skills and attitudes of those who manage them and work in them. Training has a vital role to play in achieving the objective of producing the right goods and services at the right price and marketing them successfully'.

These examples demonstrate that there is a strong relationship between the multi-skilling of employees and productivity. Within the model of the flexible firm, the multi-skilling of core employees will enable employers to deploy labour across the different occupational boundaries of the firm, as the internal conditions of production and the external demands of the market change. The cost to the employer of creating a pro-active and efficient workforce will be the investment in employee training. As the NIESR and other examples have shown, the costs of a multi-skilled workforce will be outweighed by the benefits in terms of quality, productivity and competitiveness.

Employee characteristics

The overall aptitude of employees was assessed through three measures, namely educational attainment; the nature of training; and the incidence of training. Each of these are now outlined.

Educational attainment of employees

While it may be argued that it is not educational competences which are important in the ability of employees to perform productive tasks, previous research has demonstrated that such has an important effect on the competences of workers. For example, Prais and Wagner (1985:68) through their comparative study of the educational standards within Germany and England, found that the educational basis of the workers was the main determinant of the strength of German manufacturing.

The Office of Technology Assessment has stated that; 'Educational levels are seen as an important indicator of receptiveness to learning when workplace technologies are changing rapidly; better educated workers are viewed as more likely to know how to learn.' (OTA, 1990:10). Education provides the foundation upon which the skills and aptitudes of employees can be built. For this reason the comparative levels of educational qualifications held by the work forces within the matched plants were examined.

A number of disparities are evident with regard to the comparative educational qualifications of employees within the sample of engineering firms (Table 5.1). While the Northern Ireland firms had a higher percentage of employees having completed non-certificate courses in some aspect of engineering, the Northern Ireland firms considerably trailed behind Massachusetts in the percentage of employees having completed certificate courses; BTEC or the equivalent Associate degree; Baccalaureate degrees; Masters degrees and PhD's.

Analysis of qualifications amongst the engineering employees, including those employees without any formal qualifications magnifies the disparity between the two regions, with the overall skill levels of the Northern Ireland employees appearing to be even lower relative to their Massachusetts counterparts (Table 5.1).

Table 5.1
The distribution of qualifications amongst engineering employees

	Northern Ireland	**Massachusetts**
	Percentage of employees	
No Formal Qualifications	54	28.5
Courses	10	7.5
Certificate	7	26
Associate/BTEC	15	20
Degree	13	14
Masters	1	3
PhD	0	1
Total	100	100

Source: Matched plant survey

Fifty-four per cent of the employees within the sample of engineering firms in Northern Ireland had no formal qualifications having entered the workforce directly after leaving full-time education (mostly at age 16). Not only is this figure disturbing, but the realization that these employees' levels of educational attainment had not increased since joining the firms, that is, they had not received any formal training in the form of off-the-job courses, workshops or seminars is critical to the performance of the firms.[1] A similar pattern testifying to a skills gap as measured by educational qualifications was also discovered by Roper and Hofmann (1993:48) who found that in 1989/90 only 57.6 per cent of young people in Northern Ireland left school with sufficient qualifications to start an apprenticeship as compared to 93.6 per cent in Germany.

Within the sample of engineering firms, the Massachusetts employees displayed a much higher level of overall educational attainment, while a significantly lower percentage of its workforce were without any formal qualifications than that found within the Northern Ireland engineering firms.

Table 5.2
The distribution of qualifications amongst food processing employees

	Northern Ireland	**Massachusetts**
	Percentage of employees	
No Formal Qualifications	85	79
701 / 702 / License	4	4
HNC / Certificate	3	5
HND / Associate	4	7
Degree	4	5
Total	100	100

Source: Matched plant survey

As for the engineering firms, employees within the Massachusetts food processing firms out-performed the Northern Ireland firms in the level of educational qualifications (Table 5.2). As the firms were matched by employment levels, these results emphasize that the Massachusetts employees were more likely to have higher educational qualifications than food processing employees within the Northern Ireland firms.

A higher incidence of employees who had no formal qualifications was found within the food processing sector than the engineering sector. This is however a reflection of the technical requirements of the industry, in that the food processing sector is dominated by low-skilled and repetitive work practices which do not require high levels of technical competence from the operative employees.

The overall skills base demonstrates that the Massachusetts firms are clearly at an advantage in that their employees have a higher educational aptitude. The direct implication of the educational gap upon producer quality and productivity may however be modified where a more active training policy is pursued by the Northern Ireland firms.

Attitudes to training

The 'flexible firm' model differentiates between core and peripheral workers in terms of skills levels and the investments which are made in training. The ability of workers to be functionally flexible is dependent not only upon their educational attainment, but also upon the training which they receive. The benefits to be accrued from a more flexible and efficient workforce (as measured by improved asset efficiency; return on sales; added value and profit per employee), must therefore be weighed against the costs of training. Yet it would be expected that where a firm is embracing the principles of flexible production, this would be mirrored in a willingness to invest in the training of employees, as the benefit outweighs the cost.

The attitude of managing directors towards training within the firms was ascertained from the interviews. It would be anticipated that where new market demands, new technology or new forms of production organization, were being introduced within a firm, capital and managerial investments would be mirrored by a commitment to employee training. The extent to which the managing directors perceived training as a financial burden, or as a long-term investment was therefore ascertained. From this it was possible to determine whether training was performed merely in response to job-specific skill shortages or whether training was an integral part of corporate development.

A clear difference was found between the Northern Ireland and Massachusetts engineering firms, in their attitude to training. Sixty three per cent of the Northern Ireland firms perceived employee training as a financial burden, with the remaining 37 per cent identifying it as a long-term investment. Conversely, within Massachusetts only 16.6 per cent of the engineering firms regarded training as a financial burden, with 83.4 per cent identifying the benefits to be accrued from employee training.

Within the food processing sector, again the Northern Ireland firms viewed training as a financial burden, with the Massachusetts firms perceiving it as a long-term investment. Within both regions, food processing firms saw training as more of a financial burden than amongst the engineering firms. While only 1 of the 15 Northern Ireland food processing firms perceived the potential rewards of such investment, 11 of the 15 Massachusetts firms discerned the inherent rewards of such. Through speaking to the interviewees, it was apparent that Northern Ireland firms saw training as something which 'had to be done' to enable the workers to perform their tasks, whereas generally, the Massachusetts managers could see broader implications beyond that of the mere performance of routine tasks. For example, the manager of a firm which produced various forms of cooking oils within Massachusetts, stated that:

> 'The policy of our company towards training is based on the premise that every employee knows absolutely nothing about the processes and practices of production when they are first employed. Each new-start therefore, irrespective of where they will be working in the company, undertakes the same form of induction training to become acquainted with the 'whole' production process, from the administrative aspects of the office (in terms of receipt of orders) through to the production line, to the packaging stage, and finally to the point where the product leaves the factory. By doing this, each person can see that they are an important and essential part of the overall process and we believe that this encourages them to be more diligent and conscientious in their work...in terms of the quality of their work and the enthusiasm with which they do their work' (*Interview 30/7/93*).

A discernible and fundamental difference existed in the attitudes and perceptions of the role of training within the matched plants. IFF Research Ltd (1985) also found that training costs were perceived as overheads and not as investments in the attainment of universal corporate objectives. Yet, what are the reasons for the observed views of employers? A number of points which may be described as contributory factors to the universal perceptions of managers within both regions towards training are now outlined.[2]

Within Northern Ireland at least five factors may be effecting the overall negative perception of 'training'.

1. Turnover within the Northern Ireland firms was lower than in Massachusetts, therefore employers may have a lower profit margin with which the 'luxuries' of manufacturing may be achieved (as stated by one owner of a small engineering business).
2. Northern Ireland products are highly dependent upon local and national demand (see Chapter Seven). Northern Ireland is also described as a 'sheltered' market, with low levels of import penetration. The quality of products may not be as high therefore, due to the lack of alternative products or components with which to compare standards.
3. Within the localized Northern Ireland market, there are a smaller number of 'competitor' firms than that found within Massachusetts (as identified

by the matched plants), and therefore, less incentive to develop and initiate changes in the production process, both in terms of labour and capital resources.

4. Within the Northern Ireland firms the survey found that there was less of a desire for 'growth', as the primary objective of the firm.
5. Northern Ireland firms act more as 'assemblers' and 'distributors' of manufactured goods, than their matched plants, (Hewitt, 1993:17) and therefore the need for more highly skilled staff is reduced.

Within Northern Ireland a negative attitude towards participating in training was also found amongst employees. A number of employers stated that they had established links with colleges and identified courses which would be beneficial, both to the employees concerned in terms of their career advancement, as well as to the overall performance of the firm. Yet, despite the willingness of firms to finance the relevant courses, pay for examination fees and grant employees a bonus on passing the exams, employees refused to take part in such training[3]. For example, the manager of an engineering firm in Northern Ireland stated;

> We see training as being important for the advancement of the company. Through the help of the Training and Employment Agency we established some courses which the people on the shopfloor could go to, and we even offered to pay the fees for them. If they attended the courses and passed the exams, we said that we would review their position in the company both in terms of their pay, and the possibility of promotion. But because the courses took place in the evenings they would not go....One worker said that he was tired after a full day's work and it just wasn't worth the effort to go on the course. (*Interview 18/3/93*)

This firm had offered all of its employees the opportunity to attend various courses over a six year period yet, only two people had attended and passed the examinations. Consequently, the firm highlighted that they ascertained major problems in improving the skills potential of its workers.

Morris (1988:308) also found that one of the worst problems which may be faced by firms attempting to introduce functional flexibility is the attitudes of the workers. Hence, many small firms encounter substantial obstacles in their pursuit of multi-skilling, for it is these businesses which have the bare essentials of human resources, and with the continuous pressure to meet order dates, the opportunities for either in-house or off-house training are minimized. These reasons therefore help to explain the general perception amongst the Northern Ireland firms of training as a financial burden.

It is possible to ascertain links between the views of some educators within Massachusetts of the educational courses held within Northern Ireland, and the attitude of employers to training. A number of educators referred to the Northern Ireland engineering curriculum content as producing 'trouble-shooters', as opposed to the 'innovative' engineers which they believed their courses produced. Of the Northern Ireland firms visited, only 37 per cent viewed training as being a long-term investment, and consequently it can be interpreted from such, that only 37 per cent could envisage future profitability gains from the development of their labour resources. The 'other side of the coin' to this, is that Northern Ireland employers may in reality, view their workers as 'trouble-shooters' and hence, are incapable of ascertaining the potential benefits of training upon the workplace as a whole.

In contrast to the Northern Ireland firms, the Massachusetts employers were more conscious of the relationship between training and performance. This was due to a number of factors, for example:

1. Within Massachusetts there was greater direct competition amongst firms, i.e. larger number of firms in direct competition.
2. There was a higher union presence within Massachusetts a factor which may manifest itself in a higher incidence of contractual demands serving the needs of their employees in terms of skills enhancement.
3. There was a greater aspiration for 'growth' within the Massachusetts firms, and as this can only be achieved through the creation of new products or more efficient production processes, then the development of human resources in terms of increasing the efficiency, innovativeness, quality consciousness etc. of employees is essential.
4. Within the whole culture of the Massachusetts region, there is an intense desire for advancement and the acquisition of knowledge which is largely due to the economic and educational trajectory of the state - through the cooperation of education, industry and Government - in stimulating the Research and Development base of the region.

Having outlined the predominant attitudes to training within the two regions, to what extent were these attitudes reflected in the actual training which was undertaken by the firms?

The incidence of training

A distinct difference was apparent between the Northern Ireland and Massachusetts firms in their attitude towards training. In terms of a cost-benefit analysis the Northern Ireland firms perceived training more as a financial burden than a long-term investment. Conversely, the majority of the Massachusetts firms regarded the costs of training as an investment, with the returns being manifest in improved corporate performance. A positive relationship would therefore be expected between the attitude towards training, and the subsequent investment in training by the firms. Measuring training investment in financial terms suggests the existence of a regional disparity between the matched plants. (Table 5.3).[4]

Although the Massachusetts employees had an overall higher level of educational attainment than the Northern Ireland employees, the small engineering firms within Massachusetts also spent more on training their work forces.

The Massachusetts firms were not spending more money on the training of their employees as a result of them having a lower educational and skill base, but rather they were seeking to expand upon their educational aptitudes. Conversely, within the Northern Ireland sample, while the educational attainments were lower than that of Massachusetts, the small firms were not attempting to rectify this situation, but instead, this inferior position was reinforced through lower levels of investment in employee training.

Table 5.3
Average training expenditure per firm on employees by occupational group[1]

	Shop floor	Foremen/ supervisors	Management	Total
	(£)	*(£)*	*(£)*	*(£)*
Engineering				
Northern Ireland	1650	500	985	3135
Massachusetts	3600	650	2232	6482
Food Processing				
Northern Ireland	187 **		70	257
Massachusetts	2100 **		2320	4420

Notes: **Average training expenditure per firm for supervisors is incorporated within expenditure on shop floor workers.
[1]Expenditure is measured in £ Sterling and is based upon exchange rate of £1:$1.50

Source: Matched plant survey

Within Massachusetts significantly more money was being invested in the training of employees, with both managerial and non-managerial staff attending more courses than employees within Northern Ireland. The CEST (1993:7) states that 'successful innovation is not a question of excellence in any one particular area, but overall competence throughout the firm.' In this respect, Massachusetts firms had attained the correct balance, in that they directed resources toward the whole workforce, in a manner which would increase the overall skills level of the firms, and thus coordinated development and expansion could occur. Where training is focused predominately upon one group of workers, this can result in disjointed growth, as advances within one section are not reciprocated throughout the firm, thus hindering the economic advancement of the whole firm.

Another important factor to note from Table 5.3 is the relatively low amount of money which was spent on the training of supervisory workers within firms in Massachusetts. Although it would be tempting to interpret this as a neglect of the training of supervisors within Massachusetts, this would be fallacious. Rather, results from the study show that there were 19 per cent fewer supervisors in the engineering sample within Massachusetts.

The presence of substantially fewer supervisors within Massachusetts is a direct reflection of the introduction of functionally flexible work-practices within the Massachusetts firms, as the hierarchy is reduced, and a blending of tasks occurs. An anomaly exists however in that, theoretically, flexible firms should seek to advance the skills of their 'core' workers, while exploiting those of the 'peripheral' employees (Atkinson, 1985). Yet this was not found within Massachusetts, and instead resources were directed towards all sectors of the workforce in an attempt to develop the overall skill levels within the firm. However, if closer analysis is made of the investments being made in training, then disparities *are* apparent when non-managerial and managerial staff are compared, as demonstrated by the average amount of money invested in training on a per capita basis (Table 5.4).

Regional disparities are reinforced when analysis is made of per capita, as opposed to corporate expenditure on training. Whilst average training expenditure by occupational group (Table 5.3) would suggest that the investment in managerial and non-managerial workers was similar within the Massachusetts firms, these figures conceal per capita differences in training expenditure. Instead per capita training expenditure would support the existence of a core-periphery workforce, with the multi-skilling of the core workers being coupled with a relatively lower level of training amongst the peripheral workers.

Table 5.4
Average training expenditure per employee by occupational group and region

	Shop Floor	Foremen/ supervisors	Management
	(£)	*(£)*	*(£)*
Engineering			
Northern Ireland	75	250	305
Massachusetts	180	210	720

Source: Matched plant survey

While the engineering firms within Massachusetts were investing to a greater extent in their core workers (i.e. management), this was not to the neglect and marginalization of the peripheral workers through de-skilling. Rather, the Massachusetts firms also invested a disproportionate amount of money in the training of these 'peripheral' workers as compared to the Northern Ireland matched plants.[5]

Amongst the food processing firms, again a large average difference was evident between Northern Ireland and Massachusetts firms, in the expenditure on employee training (Table 5.5). Northern Ireland firms on average spent £257 upon the training of employees, whereas the corresponding figure for the Massachusetts food processing firms was $4,420. These values however present a distorted picture of the actual expenditure upon employee training. For example, only 33.3 per cent, or 5 of 15 Northern Ireland food processing firms stated that they had spent any money on the training of employees over the 12 months prior to the interview. Amongst the Massachusetts sample, the corresponding figure was 73.3 per cent of the sample, i.e. 11 of the 15 firms. Adjusting the results to account only for those firms which participated in training produces an average corporate expenditure level amongst Northern Ireland firms of £771, as compared to $6,027 for the Massachusetts firms. Although the differential is reduced slightly, the magnitude of the difference is still sufficiently large. Yet, as was found for the engineering matched plants, again the Massachusetts firms are seen to invest much more heavily in the training of their employees, at each level of the firm, than their Northern Ireland counterparts.

Summary

The matched plant results demonstrate that employees within the Massachusetts firms were more highly qualified than their Northern Ireland counterparts. While a sectoral disparity was found in that, slightly more food processing employees had no formal qualifications within both regions, there was a clear difference between the regions with Massachusetts employees having, overall, higher educational qualifications than the Northern Ireland employees. Despite the comparative advantage in employee aptitudes, Massachusetts employers predominantly viewed training as a long-term investment as compared to the Northern Ireland perception of training as a financial burden. These perceptions of training were reflected in the expenditure on training by firms within both regions. Massachusetts firms devoted more money towards training than the Northern Ireland firms. Using expenditure on training as an indication of the presence of a core-periphery workforce (Tables 5.3 and 5..4) suggests that Northern Ireland firms had an overall low level of training expenditure and this was evenly spread between managerial and non-managerial employees. In contrast, the Massachusetts firms devoted significantly more money to the training of managerial as opposed to non-managerial workers. While this would suggest that a core-periphery workforce was being developed, comparatively larger investments in non-managerial training relative to the Northern Ireland firms suggests that the Massachusetts firms, while acknowledging the greater return from managerial training, were not polarizing the non-managerial workers, but instead, were attempting to improve skill levels throughout the firm.

Functional flexibility on the shopfloor

In examining the evidence for functional flexibility amongst shopfloor workers, a departure is found between the post-Fordist theories, in their perceived role of shopfloor workers. Within the flexible firm model the dualism in the role of labour within firms creates problems in bridging the divide between a numerically flexible periphery and a functionally flexible core. The peripheral workers are those employees on the shopfloor with low skills and low levels of training and job insecurity. In contrast the core workers, which include managers and other employees for which the supply of alternative labour in the open market is low, are multi-skilled, highly trained and secure in their employment (Atkinson, 1985). While the model acknowledges the need for multi-skilled employees within the flexible firm, this does not include the peripheral workers.

In contrast, the Flexible Specializationists and Post-Fordists propose that multi-skilling will occur throughout the flexible firm (Piore and Sabel, 1984;

Sayer, 1985; Boyer, 1988; Jessop et al., 1988). The Flexible Specializationists see new technology as stimulating organizational innovations, which demand that labour control moves away from that associated with Taylorist principles of production to one which is governed by 'autonomously self-regulating work groups' (Burrows et al., 1992:4). The Post-Fordists while being less technologically deterministic in the way they view the flexible organization of production, also state that within the flexible firm, employees will move from one job to another with fewer legal constraints within their contracts (Boyer, 1988).

It would therefore be anticipated that within the flexible firm, employees would have a low level of job demarcation; be multi-skilled and efficient at a number of tasks; and interact continuously with other employees in the organization of production (Moulaert and Swyngedouw, 1989). It should be noted however that within the context of the flexible firm model (Atkinson, 1985), functional flexibility amongst shopfloor workers may not necessarily be expected, assuming that core workers are functionally flexible.

The remainder of this section examines the extent to which functional flexibility was present amongst the shopfloor workers within the sample of small engineering and food processing firms in Northern Ireland and Massachusetts. The indicators of functional flexibility used, follow Moulaert and Swyngedouw (1989). The extent to which job demarcations are being eliminated within the firms is examined first. The evidence for multi-skilling amongst shopfloor workers within the matched plants is then presented within the context of employee interaction.

Demarcation

The benefits from implementing a policy of non-demarcation of tasks - where the employees are efficient at a variety of tasks - is unquestionable, as is measured by its effects upon the efficiency, and performance. Indeed, such a policy should also assist in the establishment of quality circles and total quality management schemes, thus improving the quality of the products being produced.

Where job demarcations are not incorporated within employee contracts it would be anticipated that this would result in greater movement of employees throughout the shopfloor. However, a precondition for employees crossing occupational boundaries is that they have the necessary skills to ensure that multi-skilling results in productivity gains as opposed to increased inefficiencies.

The Massachusetts employees had a higher overall level of educational qualifications, along with higher investments in their training. It would therefore be anticipated that the incidence of job demarcations would be lower amongst the Massachusetts firms. This assumes that those firms which had higher employee skills and had made larger investments in training, were also organizationally flexible.

Within the engineering sector, 50 per cent of the Northern Ireland firms operated a system whereby the demarcation of tasks for employees was kept to a minimum. However, within Massachusetts, 90 per cent of the firms stated that they operated a policy of non-demarcation. Conversely, amongst the food processing firms, no significant difference was recorded between the two regions, with nine of the 15 firms within both Northern Ireland and Massachusetts indicating that there was a clear demarcation of tasks on the shopfloor, with the remaining 6 firms within both samples stating that non-demarcation was *actively encouraged.*

The results would indicate that the propensity of firms to introduce non-demarcation of tasks is sectorally specific. This therefore raises questions concerning the viability of some traditional industries adopting a policy of 'flexible' production. For example, within the food processing industry the organizational principles of production have not altered very much from the principles outlined by Taylor and operationalized by Ford. Therefore, within the context of the food processing matched plants, flexible working practices were less widely used than traditional methods of labour control and organization.

Multiple tasks and employee interaction

The manner by which workers conduct their daily tasks - that is, independently, or interdependently as part of a team - provides an understanding of the extent to which shopfloor workers can perform a variety of tasks throughout the firm. This measure of functional flexibility identifies the extent of organizational flexibility within the firm. While the demarcation or non-demarcation of tasks alludes to the potential for employees to be deployed through the shopfloor, it does not automatically determine the nature of the interaction between workers. Indeed, it would be possible for firms to have a policy of non-demarcation, with workers actively crossing narrow occupational boundaries, yet still operate independently of other workers as opposed to working within a team.

The Flexible Specializationists propose that flexible technology will automatically lead to organizational innovations in the formation of independent work teams (Sayer, 1985; Murray, 1988, 1989). It would be anticipated therefore, that those firms which were pursuing organizational flexibility and the introduction of flexible production methods would also have a high level of interdependence amongst shopfloor employees, as manifest in the formation of work teams and quality control circles.

From the interviews the nature of employee interaction within both the engineering and food processing firms was ascertained (Table 5.5) in terms of extent to which employees worked independently or inter-dependently as part of a team of quality circle.

Table 5.5
Employee interaction

	Independent	Inter-dependent	Inter- and Independent	Total
	Percentage	*Percentage*	*Percentage*	*Percentage*
Engineering				
Northern Ireland	51	42	7	100
Massachusetts	11	89	0	100
Food Processing				
Northern Ireland	81	19	0	100
Massachusetts	59	41	0	100

Notes: The dominant form of employee interaction within the firms was identified as either 'independent' or 'inter-dependent'. Independent working was characterized by production operatives working on pre-determined (and often standardized) tasks, with accountability to the foremen or management. Inter-dependent working was characterized by work teams, wherein discussion and negotiation were central and employees were involved in aspects of production beyond their immediate occupational boundaries.

Source: Matched plant survey

The results (Table 5.5), demonstrate a disparity between the matched plants in the nature of shopfloor interaction. These results for employee interaction almost replicate those for the existence of job demarcations. For the Massachusetts engineering firms, those firms which operated the non-demarcation of tasks, were the same firms whose employees interacted with other workers on a daily basis.

A similar correspondence between demarcation and interaction of employees is also found for the Northern Ireland firms. Fifty per cent of the small engineering firms within Northern Ireland operated a system of demarcation, and the same 50 per cent also stated that their employees were working independently of each other. Further, although 50 per cent of the Northern Ireland firms also operated a system of non-demarcation, only 43 per cent referred to their employees' work practices as being interdependent, with the remainder stating that only some employees worked closely with other skill groups.

Within the Northern Ireland food processing firms workers mainly operated independently of each other. Combined with a high degree of demarcation of tasks this culminated in the classic Fordist organization of production with the speed of work being dictated by the assembly line and employees acting in a subservient manner to the machines.

In contrast, within the Massachusetts food processing firms there was a much wider adoption of work practices whereby employees behaved in an interdependent fashion with each other. This can not be explained in terms of 'industry specific' factors as within Northern Ireland, but rather is a reflection of the widespread formation of quality circles, designed to reduce the reject rate of production. From visiting the firms, it was apparent that within Massachusetts firms the shopfloor employees were given much greater freedom in the functions which they performed. It was not uncommon to see three or four shopfloor workers congregated around a production plan or a machine, discussing the most suitable methods of production and 'together' inspecting the work as it progressed.

Conversely within the Northen Ireland firms, such co-operation was not apparent in the working practices, with one manager stating:

> There are no benefits to be achieved from encouraging employees to work together unnecessarily. Of course there are cases where it is essential that a couple of men have to help each other, but we have found that if you let two men part-time, do a job which one man could do full-time, then it takes the two men longer...they talk about non-work related things and get distracted from their work, which can lead to more mistakes being made instead of fewer *(Interview 16/3/93).*

The greater use of quality circles within the Massachusetts engineering and food processing firms and the subsequent implications of these upon the reject rate of production, is examined later within the context of how the skill levels of the employees and the functional flexibility which this incorporates, alters the inherent competitiveness of the firms (Chapter Seven). Within the Massachusetts firms employees were encouraged to 'bond together' (*Interview 28/7/93*), in the way which work was structured, with the bonding penetrating the artificial barriers of hierarchical structures.

Employees within both engineering and food processing firms in Massachusetts were also found to be more efficient in maintaining the machinery with which they worked. For example, a number of Massachusetts interviewees highlighted that when new capital was purchased, all employees were provided with a working knowledge of the technology while a large proportion of such were also taught how to maintain the machinery. Conversely within the Northern Ireland firms, a more characteristic scenario was one where new capital was purchased and in the training course provided by the suppliers only those employees who would be in direct contact with the capital were provided with a working knowledge of it, with very little attention being paid to maintenance tasks. A study carried out by the Electronics Economic Development Council (EDC) Employment and Technology Tasks Force (as quoted in Foster, 1988) also found that many British companies failed to involve or consult their employees when planning the introduction of new technology. This observation is verified with respect to the results attained from the sample of small firms relating to the incidence of technological breakdowns and suggests that a multi-skilling policy is positively associated with both a lower reject rate of production and a smaller amount of down-time due to technical breakdowns (see Chapter Seven).[6]

Summary

The findings from the matched plants suggest that there is a significant difference in the extent to which the Northern Ireland and Massachusetts firms have implemented functional flexibility amongst shopfloor workers. For example, Massachusetts firms had a much higher incidence of non-demarcation of tasks within the engineering sector (and a comparable level of non-demarcation within the food processing sector) along with a disproportionately greater presence of workers operating interdependently - within both the engineering and food processing sectors - as opposed to the clear demarcation and non-overlapping of work tasks which was found amongst the Northern Ireland firms. Additionally, within the Massachusetts firms, this interdependence and non-demarcation of tasks culminated in the shopfloor workers being able to carry out a greater number of shopfloor functions than their Northern Ireland counterparts. Shopfloor workers within the Massachusetts firms also had a broader technical knowledge, related to the maintenance and up-keep of capital equipment. Consequently, the Massachusetts employees displayed evidence of multi-skilling to a much greater extent than the Northern Ireland employees. This multi-skilling was beneficial to the firms in that it led to a reduced number of capital breakdowns; a lower reject rate of production; and subsequently, less production time wasted through down-time (see Chapter Seven).

Functional flexibility of supervisory workers

While the flexible firm model presents a clear duality between high-skilled core workers and low-skilled peripheral workers, the role of supervisors or middle managers is less clearly defined. Yet, as for core and peripheral workers, the position of supervisors within the core or periphery of the flexible firm is controlled by the ease with which their skills can be replaced. It has been noted that within recent years the role of supervisors had been changing from that of functional specialists to generalists (Fenton-O'Creevy et al., 1994). Within the context of the flexible firm model, this would suggest that supervisors with generalist skills can be replaced, relatively easily from the external labour market and therefore they are being peripheralized within the labour force as their employment becomes less secure.

Traditionally, supervisors or foremen have acted on behalf of managers in translating policies from managers to shopfloor workers. In doing this they also monitored the performance of shopfloor workers and assessed their work. From the Flexible Specializationist perspective however, this traditional role of labour control has been radically altered in the flexible organization of production. With the introduction of new flexible technology, an environment has developed wherein, 'management itself ... should be interested in diminishing the division of labour, in reintegrating tasks and functions and improving the quality of work ... the interests of management and workers are no longer necessarily antagonistic' (Dankbaar, 1988:26).

Flexible technology demands organizational innovations which are 'predicated on collaboration (Piore and Sabel, 1984:278). Greater employee involvement is therefore seen as inevitable with employee control and determination (Sayer, 1985; Piore and Sable, 1984) being positively associated with 'increased job satisfaction for employees, and increased commitment to the organization, greater success in managing change, more effective use of tacit knowledge, lower turnover and absenteeism and in some cases significantly improved productivity' (Fenton-O'Creevy et al., 1994: I).

This section examines the role of supervisors within the engineering and food processing matched plants in Northern Ireland and Massachusetts. The extent to which supervisors are functionally flexible is gauged by the incidence of supervisors within the small firms; the number of employees for which supervisors are responsible; the nature of supervisory skills; the nature and incidence of training; the skills acquired through training; and the ability of supervisors to perform specialist functions within the firm.

The number of supervisors

Within Fordism supervisors were of prime significance in monitoring production and ensuring the efficient running of the assembly line. However, under flexibility, it would be expected that firms undergoing the rationalization and restructuring of production would eliminate the 'hierarchical' structure and replace it with increased functional flexibility amongst employees. Under these new principles of production workers can work interdependently and through 'teamwork', create an environment whereby discussion and innovation are encouraged. Providing employees with greater involvement and responsibility also nurtures the less quantifiable skills of firm loyalty, enthusiasm and conscientiousness in their work (Fenton-O'Creevy et al., 1994).

Within the Massachusetts engineering firms, 19 per cent fewer supervisors were found than within the Northern Ireland matched plants. This illustrates that within the Massachusetts sample, the traditional managerial hierarchy of control is being refined through a blending of tasks and occupational boundaries. A comparatively lower number of supervisors within the Massachusetts plants in conjunction with the higher skill levels of employees, suggests that the need for continual monitoring of production, and subsequently the number of supervisors within the production process, is reduced.

With respect to the sample of food processing firms a similar number of supervisors was found within both the Northern Ireland and Massachusetts firms. This suggests that within the food processing sector Fordist techniques of production are still actively pursued, as manifest through the maintenance of traditional hierarchical structures of control. It also suggests that collaboration and the minimization of labour control may be more difficult in some sectors than others.

In terms of the number of shopfloor workers for which supervisors were responsible, again, differences were apparent between the matched plants. Amongst the Northern Ireland engineering firms, each supervisor was responsible for 7.6 employees while the equivalent figure for Massachusetts was 8.6 employees per supervisor.

Although the difference between the matched plants is slight, it should be remembered that the overall employment level within these firms is small. Where supervisors are an integral part of labour control within the firm, then the ratio of non-management to supervisory workers would indicate a more efficient utilization of supervisors within Massachusetts. The higher worker to supervisory ratio, within the Massachusetts engineering firms, suggests that these firms are a step closer to attaining a flexible workforce - as monitoring and checking of work, (which is traditionally the domain of the supervisor) is reduced and workers are encouraged to oversee the quality of their own work.

Amongst the food processing firms, again the Massachusetts supervisors were responsible for a larger number of employees than supervisors within the Northern Ireland firms, however the difference was slight. Within the Northern Ireland firms, on average each supervisor was responsible for 10.2 employees with the corresponding figure for the Massachusetts firms being 11 employees. This however conceals the greater use within the Northern Ireland firms of charge-hands, who were responsible for labour control and the flow of production along the assembly line. Each charge-hand, while acting in a traditional supervisory role, also performed standardized tasks. Charge-hands by virtue of their skills were peripheral within the workforce, in that they could easily be replaced by labour from the open market.

Charge-hands were not found within the Massachusetts food processing firms to the same extent as within Northern Ireland. Instead, the monitoring of production flow and quality checks were functions which the production operatives performed. Quality circles were more important within the Massachusetts firms, as evident through greater worker interdependence and collaboration (see Table 5.5). Through the interviews conducted with managers of small food processing firms in Massachusetts the desire of such employers to achieve a high level of quality control was apparent throughout. Indeed many employers pointed out that they had adopted a policy of Total Quality Management and the introduction of Quality circles within the firm to reduce the overall reject rate and associated inefficiencies. Where employees were encouraged to work together and monitor the quality of their own production, the need for supervisors was reduced.

Supervisory skills

Flexible Specializationists and post-Fordists within the Regulation school of thought propose that an overall upgrading of skills will occur throughout the workplace. Recent research has suggested that supervisory skills are becoming more generalized with a greater need for people management skills and the 'lateral integration of work processes' (Fenton-O'Creevy et al., 1994:I). Within the flexible firm it would be anticipated that if supervisors are part of the core workforce then their skills would enable them to cross job and occupational boundaries. However, if supervisors are mainly low-skilled workers, and they are acting in a labour control capacity, then they will be part of the peripheral workforce.

In ascertaining whether or not supervisors were part of the core or periphery workforce, the extent of multi-skilling amongst supervisory workers was examined. As supervisors are mainly promoted from the shopfloor within small firms[7], the managing directors were asked to identify which skills were essential in appointing a supervisor. Despite this variable being highly subjective, it is useful in determining the extent to which collaboration as opposed to control existed within the firms. The responses attained by managers relating to the skills that they required their supervisors to possess were sub-divided into seven categories, namely:

> *Technical skills* related to the knowledge of machinery (including both setting and operation) and the ability to overcome any technical problems that should arise.
>
> *Efficiency skills,* where supervisors could detect where production methods could be made more efficient.

Worker related skills, in terms of the ability of supervisors to communicate with shopfloor workers and establish a positive relationship through which managers could introduce changes within the production process with the minimum opposition from employees.

Dependability characteristics, that symbolize both commitment and loyalty to the firm and are manifest through conscientious work practices and the close monitoring of other employees' performance.

Flexibility skills, which enable the supervisors to be adaptable across occupational boundaries, which although being predominantly related to production skills, may also encompass the administrative and 'business skills' aspect of the firm.

Quality related skills, that are manifest in the close monitoring of production at various stages of the production process and demand both accuracy and precision.

Innovative skills, that enable the supervisors to identify how the production process could be altered to produce a better quality of product, new products, or new methods of production.

By means of this categorization the relative importance of each of the skill groups was ascertained (Table 5.6).

Within Massachusetts only 16 of the 35 engineering firms had supervisors as compared to 27 within Northern Ireland. Within the Northern Ireland firms appointment of supervisors depended upon their technical aptitudes along with their ability to motivate, communicate and deal with other workers. In contrast, the Massachusetts engineering firms predominantly appointed supervisors on their ability to be adaptable, or flexible across occupational boundaries.

These results are indicative of the differences between the matched plants in the organization of production. For example, the Northern Ireland firms demonstrate a rigidity in their structures of control and management, with supervisors being required to be technically competent at a range of tasks, acting in many cases as 'trouble-shooters'. Supervisors within the Northern Ireland engineering firms were required to have an amiable relationship with the other workers, and therefore provide a 'calming' influence upon the introduction of new work practices, minimizing the antagonism between management and workers. Further, the engineering supervisors within Northern Ireland were required to inspect and monitor production both in terms of the quality of the products and the conscientiousness of shopfloor workers.

The extent to which supervisors within the Massachusetts engineering firms acted as intermediaries and inspectors was considerably less than within

Northern Ireland. As previously outlined, the interdependent and group relations established through the adoption of quality circles ensures continuous monitoring of production and the reduction of rest periods due to the necessity of being part of a team. Additionally, within this more flexible structure (including the elimination of middle management), worker attributes of trust, conscientiousness, enthusiasm and loyalty are encouraged, and thus employees work more intensely for the overall advancement of the firm, without as great a need for labour control.

Table 5.6
The relative importance of supervisory skills[1]

	Northern Ireland	**Massachu-setts**	**Northern Ireland**	**Massachu-setts**
	Engineering (Percentage of firms)		*Food processing (Percentage of firms)*	
Technical	30	22	14	34
Efficiency	9	0	6	0
Worker Relations	35	13	20	40
Dependability	0	8	20	0
Flexibility	17	37	0	14
Quality Related	4.5	12	34	0
Innovative	4.5	8	0	6
No Specific Skill	0	0	6	6
Total	100	100	100	100

Notes: [1] Values represent the percentage of firms within the sample identifying a specific skill group.

Source: Matched plant survey

As for the engineering firms, again a number of dissimilarities were found between the food processing firms in the perceived attitudes of employers to the role of supervisors. Firstly, within Massachusetts there was a greater emphasis upon the possession of technical skills than within Northern Ireland. Secondly, the Massachusetts food processing firms also placed a much higher emphasis upon the ability of supervisors to act as 'intermediaries' between management and shopfloor workers. Thirdly, Massachusetts firms emphasized the possession of flexibility and innovative skills, yet these skills were not regarded as important for supervisors to possess within the Northern Ireland food processing firms. Finally, the Northern Ireland firms surpassed the Massachusetts firms in their desire for supervisors to hold quality related skills[8].

The results illustrate that both the engineering and food processing firms within Massachusetts demand a higher degree of 'flexibility' from their supervisors. In terms of the engineering sector, Massachusetts supervisors must be flexible and adaptable whereas the Northern Ireland firms merely seek workers who are competent in their use of capital and able to act as good communicators between management and shopfloor workers. Within the food processing sector, again Massachusetts firms demand that supervisory workers are 'flexible' and 'innovative' while also having technical skills and displaying the ability to act as intermediaries. Conversely, within the Northern Ireland food processing firms supervisors are employed as quality control operatives, a factor which accentuates their role as 'big brother' as they monitor and assess the work of production operatives (workers which within Massachusetts have responsibility for their own work, both through continuous checking and quality circles).

Nature and incidence of supervisory training

With the introduction of new, flexible technology, the Flexible Specializationists propose that the skills of all employees will be upgraded. The improvement of supervisory skills is however dependent upon training. Although supervisors have traditionally been identified as having a poor educational base and a low level of technical skills, these ineptitudes have been reinforced by limited training (Lowe, 1992). Under conditions of flexibility however, supervisors would have their skills upgraded to permit multi-tasking throughout the firm.

Traditionally, training for supervisors has been on-the-job, and therefore has been characterized as being *ad hoc*, informal and unstructured. Within the flexible firm, due to the centrality of training in equipping supervisors with multi-skills, training would be more organized, structured and targeted. The format of training would therefore move from the traditional dependence upon on-the-job *ad hoc* instruction, to structured off-the-job learning.

Northern Ireland and Massachusetts firms within each of the engineering sectors identified on-the-job training as being the most important form of training for supervisors (Table 5.7). The only exception to this was the electronics sector within Massachusetts where off-the-job training dominated. The disparity between the amount of on-the-job and off-the-job training was greatest within the Northern Ireland firms, while off-the-job training was used more extensively within the Massachusetts firms.

Table 5.7
The proportion of supervisory training which is on-the-job and off-the-job by sector

	On-The-Job	Off-The-Job	Total Training
	Percentage of Supervisors		
Northern Ireland			
Metal Products	81	19	100
Mechanical Engineering	79	21	100
Electronics	62	38	100
Food Processing	100	0	100
Massachusetts			
Metal Products	69	31	100
Mechanical Engineering	70	30	100
Electronics	56	44	100
Food Processing	87	13	100

Source: Matched plant survey

The extent to which on-the-job or off-the-job training is used is dependent upon the sector, with electronics firms within both regions being more dependent upon off-the-job training. Within the food processing sector 9 of the 15 Northern Ireland firms reported that all supervisory training was carried out on-the-job - equating to 60 per cent of the sample. Additionally, the remaining 40 per cent of these firms stated that they did not train their supervisors. The results obtained were slightly different for the Massachusetts firms, with only 20 per cent, or three of the 15 interviewed, stating that they performed 100 per cent of supervisory training on-the-job, with only five of the firms not undertaking any supervisory training.

A difference was also found between the matched plants in the extent to which supervisors were encouraged to attain further qualifications. The attainment of further qualifications are important because they go beyond the job-specific nature of non-certificate courses and training (Foster, 1988) and assist employees in obtaining more cognitive skills. While 80 per cent of the Massachusetts engineering firms stated that they encouraged their supervisors to enrol for further qualifications in colleges, only 9.5 per cent of the Northern Ireland firms encouraged this.

The Massachusetts food processing firms also differed from their Northern Ireland matched plant in that off-the-job training was used for some of the supervisors. Generally it was found that the format of off-the-job training consisted of in-house instruction related specifically to planning, innovation, and time management. However, some of the supervisors were 'licensed' in their occupations - mostly bakers - and at intervals attended courses at colleges to fulfil the requirements for the renewal of their licenses. By doing this, their skills were refreshed and they 'returned like a breath of fresh air, often bringing new ideas about products and processes' (*Interview 20/7/93*).

As was highlighted within the engineering sector, despite the Massachusetts firms using off-the-job training to a greater extent than their Northern Ireland counterparts, the real per capita investment in supervisory training is lower within Massachusetts, equating to \$347 or £231 (at an exchange rate of £1:\$1.50) with the Northern Ireland value being £249. These figures are of crucial importance in understanding the extent of functional flexibility amongst supervisors. Supervisors within Massachusetts are perceived more as peripheral workers than within the Northern Ireland firms. This peripherality is evident in two forms. Firstly, the Massachusetts firms have 19 per cent fewer supervisors within the engineering sector and a smaller absolute number of food processing supervisors, when charge-hands are included in the Northern Ireland sample. Secondly, investments in training of supervisors as compared to managers and shopfloor workers demonstrates greater marginalization within the Massachusetts firms of supervisors and shopfloor employees, relative to managers, than is found within the Northern Ireland firms.

While the Northern Ireland firms are maintaining the structures and characteristics of Fordism (as evident in a more pronounced managerial hierarchy and supervisors performing traditional labour control roles) the Massachusetts firms appear to be restructuring their work forces in a way which is eliminating the managerial hierarchy while simultaneously training supervisors to be adaptable across occupational boundaries.

Supervisory skills acquired through training

Bramlette (1984) has proposed six typologies of supervisors which may be found within firms, these are:

> *The leadman* is employed as a result of past experience and performance and has direct responsibility for production and service activities;

The one-to-one supervisor is responsible for the work of employees and therefore must be skilled in monitoring production and directing and controlling workers, from which inter-personal skills are essential;

The group leader has responsibility for the activities of a group and must be able to overcome problems which emerge;

The team coordinator manages the activities of an interdependent team of workers and ensures effective communication amongst each person;

The team boundary manager has responsibility for a number of teams and informs the different groups on the level of performance and productivity;

The team resource person acts as a mentor in assisting teams with problem solving and trouble shooting, and possesses a high level of technical knowledge.

Supervisory skills and the nature of training which supervisors receive will therefore be indicative of the role which they play within the firm.

Within the flexible firm, not only will supervisors be undergoing the upgrading of their skills, but the type of skills which they require will also change as the organization of production moves away from the traditional format of the assembly line. Skills therefore mirror the organizational structures within the firm. For example, it has traditionally been important for supervisors to possess monitoring and interpersonal skills to allow them to undertake one-to-one supervision. Within the flexible firm however, the role of supervisors moves away from that of labour control and quality monitoring to that of team coordinator or team boundary manager, for whom communication, technical knowledge and adaptability throughout the workplace is essential (Bramlette, 1984).

The relative importance of different types of supervisory skills which are acquired through training, was examined within the matched plants (Table 5.8).

Table 5.8
The type of skills acquired through training

	Northern Ireland	**Massachusetts**
Percentage of supervisors		
Technical	62	32
Supervisory	22	8
Efficiency	16	0
Knowledge	0	44
Safety	0	8
Innovative	0	8
Total	100	100

Source: Matched plant survey

The skills acquired from training highlight that within the engineering firms technical skills were central to the role of supervisors. These technical skills allowed the Northern Ireland engineering supervisors to efficiently manage and monitor production within their departments, and identify the source of any production difficulties.

Within the Massachusetts engineering firms, the acquisition and development of technical skills was surpassed by the attainment of general and industry wide knowledge which was believed to assist supervisors in moving more freely and identifying inefficiencies throughout the firm (Fenton-O'Creevy et al., 1994). There is therefore a positive association between the supervisory qualities perceived to be important by the managing directors, and the nature of the skills which supervisors acquired through training.

Within the Northern Ireland food processing sector, none of the firms visited operated a training policy for supervisors. These supervisors therefore received no formal training. As they were appointed to their job through internal promotion, it was assumed that they 'knew' their jobs and were capable of operating efficiently throughout the plant (*Interview 26/4/93*).

Similarly within Massachusetts, 80 per cent of the firms did not operate a training scheme for their supervisors. However, for the three of the 15 firms which did operate a training scheme, only two firms stated that their supervisors acquired 'technical' skills from training, while the remaining firm geared training of supervisors towards the acquisition of 'innovative' skills, i.e. the ability to identify where process and product innovations could be made.

While the total amount of training for supervisors within the food processing firms was minimal, where it did occur, and particularly within Massachusetts, a positive correlation was identifiable between the skills perceived as necessary and the skills acquired through training. Within the food processing matched plants there was a deficiency in the training of supervisors, with this being more accentuated within the Northern Ireland firms. Where supervisors act as leadmen or one-to-one supervisors, in monitoring the work of other employees and advising on production problems, it would be prudent to ensure that they were highly trained. Supervisors within this role should be capable of foreseeing problems within the production process and develop more efficient methods of production to reduce both the wastage of raw materials and maximize profitability.

The type of supervisory skills acquired through training suggest that, in line with Bramlette's typologies (1984), supervisors perform distinctly different functions. These functions are indicative of underlying differences in the organizational structure of production within the matched plants. The Northern Ireland supervisors within both the engineering and food processing firms act in a traditional Fordist role, by monitoring production, checking quality and controlling labour. Conversely, the Massachusetts engineering supervisors displayed greater flexibility in occupying a team coordinator role, whereby they managed the activities of a team of workers and sought to overcome any problems which arose.

Cross-occupational skills

Under flexible specialization not only will supervisors' skills be upgraded but the attainment of functional flexibility permits the deployment of 'core' workers throughout the firm. If the supervisors within the matched plants are part of the functionally flexible core, then they should be capable of crossing occupational boundaries within the firm.

In examining the extent to which supervisors could cross occupational boundaries five main categories of cross-occupational skills were identified including, planning; design; marketing; purchasing; and Research and Development.

Table 5.9
The percentage of engineering supervisors capable of performing additional occupational tasks

	Northern Ireland	**Massachusetts**
Percentage of supervisors		
Planning	13	72
Design	40	80
Marketing	5	13
Purchasing	5	20
R&D	5	50
No Higher Functions	58	13
Total	100	100

Source: Matched plant survey

The results (Table 5.9), illustrate that the engineering supervisors within the Massachusetts firms were capable of performing a wider range of jobs than their Northern Ireland counterparts. In particular, Massachusetts supervisors were more capable of performing planning, design and research and development activities within the firm. Supervisors within both regions were however weak in terms of their marketing and purchasing skills. Whereas, 58 per cent of the Northern Ireland supervisors were unable to perform any of the five occupational jobs, this only applied to 13 per cent of the Massachusetts supervisors.

In terms of multi-skilling, such data would suggest that supervisors within the Massachusetts firms, (where they have not been eliminated through restructuring) display a much higher propensity to perform a broader range of tasks than their Northern Ireland counterparts.

Within the Northern Ireland food processing firms none of the supervisors were capable of performing any of the additional occupational jobs. Conversely, within the Massachusetts food processing firms 53.3 per cent of the supervisors were capable of performing 'planning' functions within the firm, with the remaining supervisors also being unable to perform any of the additional jobs.

Although the majority of food processing firms within both regions, had supervisors who were unable to perform additional occupational jobs, this was more pronounced for the Northern Ireland firms. Further, a significantly higher percentage of engineering supervisors within the Massachusetts firms were capable of being deployed across a greater range of occupational boundaries. In particular, the ability of more Massachusetts supervisors to perform planning, design and research and development reinforces previous findings, that, through multi-skilling the role of these supervisors had moved away from labour control, to include both the conception of tasks (through research and development), and the execution of tasks (through production planning and the coordination of group work) (Jessop et al., 1988).

Summary

The results have demonstrated that Massachusetts firms have moved closer to a flexible workforce than the Northern Ireland firms, through the diminished presence of a middle-managerial strata in the hierarchy of the firm. Additionally, those supervisors which were present within both engineering and food processing sectors, in Massachusetts were responsible for a larger number of shopfloor workers than their Northern Ireland counterparts. This was indicative of the higher expectations within Massachusetts firms of shopfloor workers (especially within engineering) where they took on responsibility for the quality of their own work, as well as performing maintenance tasks on machinery. In such circumstances the supervisors were merely overseers, and their functions were defined by their need to be flexible and adaptable throughout the workplace, as well as having the capacity to be innovative and identify areas wherein inefficiencies could be eliminated. Conversely, within the Northern Ireland firms supervisory roles took the form of quality and labour controllers as they checked the goods being produced, and monitored the performance of shopfloor workers.

The Association of British Chamber of Commerce (ABCC) in a consultative document (as quoted in Foster, 1988) highlighted that as the skill content of jobs increases there will be greater use of off-the-job training, with the predominant requirements of training being knowledge-based, as opposed to skills-based. Indeed within the Massachusetts sample, the training which supervisors received was focused towards increasing their knowledge levels with respect to the industry and the relevant products, and this was manifest in a disproportionate amount of training being off-the-job (as compared to Northern Ireland). Alternatively, the Northern Ireland firms - both engineering and food processing - almost exclusively used on-the-job training for their supervisors, with the objectives of this being directed toward the improvement of the technical capacity of supervisors, that is, their ability to act as 'trouble-shooters' within the firm, as well as improving their quality awareness, as they monitor various stages of the production process.

Within engineering, the Massachusetts firms were restructuring their work forces and attempting to eliminate what they perceived to be the inefficiencies of middle management. Although the difference between the expenditure on the training of supervisors between Northern Ireland and Massachusetts was minimal, the nature of this training was significant. While Northern Ireland firms concentrated on the traditional functions of labour control and the cultivation of positive workers relations, the Massachusetts supervisors were trained to be innovative and adaptable throughout the workplace. Further, within Massachusetts supervisors were encouraged to improve their skills base through the attainment of further qualifications which expanded beyond the short-term specific requirements of the firm. Therefore, it can be concluded that the Massachusetts supervisors within the engineering firms were aspiring towards greater functional flexibility, through multi-skilling. Alternatively, engineering firms within Northern Ireland were entrenched within the traditional forms of production, and thus their supervisors were taking on the characteristics of supervisors within Fordist production. Although the Northern Ireland supervisors were trained, such was narrowly focused toward the specific needs of the companies and in many ways was leading to the de-skilling of such workers as technology changed and their role as appendages of the machines culminated in their purpose being defined by the monitoring of both the performance of workers and the quality of goods produced.

Within the food processing matched plants, the role which supervisors played was more comparable between the regions. The Massachusetts supervisors however, demonstrated the characteristics of functional flexibility much more than the Northern Ireland employees. For example, in the perceived role of supervisors within food processing plants, the Northern Ireland firms once again concentrated on the ability of supervisors to be quality checkers, whereas the Massachusetts firms identified technical skills, along with the need to be flexible and innovative throughout the firm. Additionally, the Massachusetts firms employed a greater use of off-the-job training, and encouraged supervisors to attain further qualifications much more than within the Northern Ireland firms. Therefore, while the food processing industry, as a result of its inherent nature, does not lend itself fully to the ideals of 'flexibility', this does not mean that it cannot benefit from the advantages of developing some aspects of functional flexibility, as illustrated by the practices of the Massachusetts firms.

Functional flexibility of managers

The perceived role of managers has changed in response to organizational and technological innovations within the firm. Fayol (1916) described the role of managers as 'to plan, organize, coordinate and control'. Since then other managerial scientists (Mintzberg, 1975; Stewart, 1988) have acknowledged that managers also encounter considerable 'task fragmentation, unscheduled demands, episodic face-to-face communication, networking, and use of informal information and extemporary thought and action' (Fenton-O'Creevy et al., 1994:9). Indeed, Papadimitriou (1986:47) places a premium on 'initiative, quick responses and diagnostic skills' within the flexible firm.

Whilst the functions and tasks which managers undertake may be a function of the size of the firm, the primary objective of all managers must be the most efficient use of available resources. Indeed, small firms may be unable to support the managerial excellence that they need, and will therefore be forced to be more versatile in the jobs which they perform. Drucker (as quoted in Hines, 1993:2) states that managers' jobs will be characterized by a three-tier classification. These tiers include 'managing a business, managing managers and managing workers and work ... but in its daily work management cannot separate them ... the specific situation of the manager [is] to have not one but three jobs at the same time.'

While traditionally managerial roles have been strongly influenced by the size of the firm, the Flexible Specializationists perceive the introduction of flexible technology as demanding a similar response from all managers. Within Fordism, the role of management demonstrated a clear distinction between the 'conception and execution' of production (Aglietta, 1979). Under conditions of flexibility, some would argue that the demarcation between conception and execution of work is becoming blurred (Jessop et al., 1988). Indeed, under flexible methods of production the need for a more knowledgeable workforce (including management) is especially strong due to the vulnerability of the system of disruptions. Sayer and Walker (1992) therefore conclude that, within the flexible firm management must become more involved on the shopfloor. 'Management's purposes include reducing the cost and time of circulation, coordinating far-flung production processes, and linking pre- and post-production labor inputs for more effective product development and sale. Information handling ... is a basic element of managerial labor' (Arrow, 1985 as quoted in Sayer and Walker, 1992:70).

This section examines the extent to which managers within the matched plants acted in a Fordist manner, that is, in solely undertaking the planning, organizing, coordinating and controlling functions within the firm, or whether they displayed greater levels of functional flexibility throughout the firm as suggested by the flexible firm model (Atkinson, 1985). The indicators used include, the number of managers within the firms; the qualifications of managers; the nature and investments in managerial training; the skills which were obtained through training and the ability of managers to undertake jobs upon the shopfloor. Each of these measures are examined in turn.

Number of managers

Within the flexible firm model the core workforce consists of highly skilled employees and managers. Within the context of the small firm however, where highly-skilled employees are present they are almost always ascribed the title 'manager'. Therefore an examination of 'managerial employees' encompasses both those managers involved in traditional planning, organizing and controlling functions as well as the professional engineers and draughters etc.

It is possible that if there was a significant difference in the number of managers employed within the matched plants between the two regions then this may contribute to the disparity in growth rates (see Chapter Four). However, results from the study illustrated that there was no significant difference in the number of managers within the two regions, with the Northern Ireland firms having 10 per cent of their workforce composed of managers as compared to 8 per cent for the Massachusetts firms.

Within the Northern Ireland firms it was found that together, managers and supervisors comprized 16 per cent of the total work force as compared to 13 per cent within the Massachusetts firms. Therefore, the Massachusetts firms had a similar proportion of managers[9], slightly fewer supervisors (or middle-managers) and slightly more shopfloor workers. Where additional managers would permit a more efficient utilization of resources within small firms, through the fragmentation of managerial functions, a comparative advantage was not found for the matched plants within either Northern Ireland or Massachusetts.

Managerial qualifications

Although the direct correlation between managerial qualifications and small firm growth has been questioned (Barkham, 1992; Woo et al., 1989; Wynarczyk et al., 1993; Storey, 1982; Dunkelberg et al., 1987; Kalleberg and Leicht, 1991; Solem and Steiner, 1989; Reynolds and Miller, 1988; and Westhead and Birley, 1993) other sources (Storey, 1994) have stressed the existence of an important relationship between the two factors. Indeed, although some commentators would have reservations concerning the nature of this relationship, it is generally accepted that the positive results obtained suggest the existence of a broad relationship between firm growth and the educational background of the manager.

Other studies however, such as that by Prais and Wagner (1985) found that the strength of the vocational training displayed by German industry originated within the schooling stage. Further the Office of Technology Assessment (OTA, 1990:10) also noted that educational levels were an important indicator of receptiveness to learning within an environment of rapidly changing technologies with better educated workers being viewed as more inclined to know how to learn.

It is therefore essential that the comparative levels of educational qualifications held by the work forces within the matched plants is examined in establishing the degree of multi-skilling and the subsequent ability of firms to display flexible labour practices. Within all of the companies visited, details of qualifications were obtained for the managers in employment. These qualifications were regarded as 'formal', that is, they led to the attainment of a certificate and could be classified as 'academic' or 'vocational'. Therefore vocational courses which did not result in certification were excluded in terms of qualifications, yet where such was undertaken within the 12 months prior to the interview, they were incorporated within the analysis of training.

The results (Table 5.10) illustrate that there was a lower percentage of higher qualifications amongst the Northern Ireland managers. Northern Ireland firms had a higher proportion of engineering managers holding BTEC qualifications as compared to the equivalent Associate degree within Massachusetts. However, although there was comparability in terms of the proportion of engineering managers having 'degrees' between both regions, the greatest disparities occurred for those holding Masters degrees and PhD's.

The qualification levels of managers within the food processing sector was slightly different from the engineering sector, with a larger proportion of the managers having no formal qualifications. This low incidence of qualifications may be due to a number of reasons, including the fact that the food processing sector is relatively low-skilled as compared to the engineering sector and therefore there is not as great a need for high educational and vocational qualifications. Additionally, the educational barriers to entry are lower within the food processing sector.

Evidence from the research suggests that managers within the Massachusetts matched plants were qualified to a higher level than their Northern Ireland counterparts. Yet it would be fallacious to judge the competencies of managers within both regions solely on the basis of formal qualifications and therefore account must be made of managerial training because it is the training and retraining of managers which is regarded as one of the most important elements in determining the future prosperity of the firm.

Table 5.10
Percentage distribution of managerial qualifications

	Engineering		Food Processing	
	Northern Ireland	*Massachusetts*	*Northern Ireland*	*Massachusetts*
	Percentage		*Percentage*	
No Formal Qualifications	10	7	37	35
Certificate/Higher National Certificate	0	0	16	0
Associate/Higher National Diploma	23	5	14	20
Degree	58	52	33	45
Masters	8	31	0	0
PhD	1	5	0	0
TOTAL	100	100	100	100

Source: Matched plant survey

Managerial Training

Managerial training has been defined as: 'the process by which managers acquire the knowledge and skills related to their work requirements by formal, structured or guided means' (Deloitte, Haskins and Sells, 1989:3). The Bolton Report (1971) highlighted that the low quality of management was a major problem facing the survival and internal efficiency of small firms in the UK (Storey and Westhead, 1994). Within the context of the matched plant study - and in light of the findings pertaining to the educational and vocational skill levels of the managers - the extent to which training took place and the nature of this training was examined.

An estimate of the amount of training which managers received can be gauged by the expenditure on training by the firms over the 12 months prior to the interviews (see Tables 5.3 and 5.4). On average Northern Ireland engineering firms spent £305 per manager as compared to £720 per manager within the Massachusetts engineering firms. By breaking the data down to account for the amount spent on off-the-job and on-the-job training the average Northern Ireland firm directed 76.8 per cent of their investment in managerial training towards off-the-job training with the reciprocal value for the Massachusetts firms being 94.8 per cent. These results therefore demonstrate that, in both absolute and relative terms, the Massachusetts engineering firms invested more in off-the-job managerial training than their Northern Ireland counterparts. The Association of British Chamber of Commerce has highlighted that as the skill content of jobs increases there will be greater use of off-the-job training (Foster, 1988). The nature of managerial training would therefore suggest that the greater use of off-the-job training amongst the engineering firms was indicative of a greater increase in the skill content of managerial jobs within Massachusetts.

Differences were also found between the matched plants in terms of managerial training within the engineering sectors. Within the metal products sector, the Northern Ireland investment in managerial training was equivalent to 117 per cent that of the Massachusetts firms. However for the mechanical engineering and electronics sectors, the Massachusetts investment equated to 144 per cent and 163 per cent that of the Northern Ireland investment, respectively. This is important for it is the later industrial classifications, and especially the electrical and electronics sector, which hold the potential for corporate growth, through a higher profit to asset ratio. Conversely, the metal products sector is characterized as a traditional industry, where there is low value-added, and profit margins are small relative to the other industrial sectors. The investment gap in managerial training was however greatest within those sectors which provided the greatest potential for growth.

Within the food processing firms, the average Northern Ireland firm had invested £70 in the training of its managers in the 12 months prior to the interview, as compared to £2320 within the Massachusetts firms (Table 5.3). While 70 per cent of training for managers within the food processing firms was off-the-job the corresponding figure for the Massachusetts managers was 91 per cent. As within the engineering matched plants, again the Massachusetts food processing firms performed a higher proportion of their training off-the-job, than Northern Ireland firms. Yet within the food processing sector the nature of training is overshadowed by the disparity between the Northern Ireland and Massachusetts firms in the absolute investment in managerial training.

Significant differences were therefore apparent between the matched plants, in the amount of money that was invested in managerial training. Without exception the Massachusetts firms had a higher per capita investment in managerial training than their Northern Ireland matched plants. Further, off-the-job training was significantly more important in the training of managers within the Massachusetts firms. Assuming a positive association between training and multi-skilling (Atkinson, 1985; 1988; Foster, 1988), these results suggest that, not only was the skill content of Massachusetts managers' jobs increasing at a greater rate than within Northern Ireland, but the higher investments in training within Massachusetts was equipping the managers with the aptitudes to perform their jobs more efficiently.

Skills acquired through training

Differences have been identified between the matched plants in terms of managerial qualifications and the amount of money which was being invested in managerial training. Yet these indicators are merely indicative of the extent to which managers are functionally flexible. The skills which managers were acquiring from training illustrates the extent to which the present skill content of managers conforms to the traditional pattern of 'conception' tasks (Fayol, 1916), or whether there is a merging of production and white-collar functions within the flexible firm (Jessop et al., 1988).

Managers within the matched plants identified a number of skill groups towards which training was targeted. These included:

> *Technical advancement skills* - that is the acquisition of knowledge directly related to the types of capital equipment used within production. Through training, the managers learnt about advanced technology and the applications of such. Most of this training occurred on machine manufacturers' premises, or occasionally, a trade association would run a seminar dealing with new types of machinery.
>
> *Product development skills* - these included courses dealing with products which were being produced, and frequently through the focusing on areas of product manufacture, ideas were generated for the managers with respect to ways to refine the products being produced, or alternatively the introduction of new products which were complementary to the existing product range.
>
> *Process development skills* - this provided managers with a knowledge of new technology and processes which were evolving and could be used in the production process. The managers generally stated that courses on process development culminated in the development and refinement of existing processes within the firm.

Knowledge related skills - such skills were attained in a variety of ways and sought to broaden the awareness of managers with respect to a variety of factors, ranging from industry specific matters, to broad economic and local conditions of economic development.

Various skills - this category was used where managers were unable to define specifically the type of skills acquired from training, and usually encroached upon aspects of the other skill groups.

Market related and business skills - this category related to the development of marketing, purchasing, financial/accounting, administrative and strategic planning skills and consisted of a variety of both in-house and off-house training.

Significant differences exist in the type of skills managers acquired through training (Table 5.11). Within the engineering sector the majority of training within the Northern Ireland firms was directed at improving the technical skills of managers. Within the Massachusetts engineering firms process development skills dominated in the training of managers. The disparity in focus of managerial training links directly with the results for the investment and format of training. For example, in Northern Ireland the prominence of technical skills in the training of managers is directly related to the nature of training, with a large percentage of training being on-the-job and dominated by machine suppliers as they demonstrate new skills associated with new machinery. Further, in contrast to the Massachusetts matched plants, the Northern Ireland firms were seen to under-invest in the training of their managers[10].

Table 5.11
Percentage distribution of skills acquired by managers

	Engineering		Food processing	
	Northern Ireland	*Massachusetts*	*Northern Ireland*	*Massachusetts*
Percentage of managers				
Technical	50	14	34	20
Product Development	11	19	0	0
Process Development	11	36	0	20
Knowledge	11	19	7	46
Various	0	8	0	0
Market & Business	17	4	7	14
N/A	0	0	52	0
Total	100	100	100	100

Note: Values represent the proportion of managers for whom training was concentrated within each of the skill groups.

Source: Matched plant survey

Within Massachusetts process development skills surpassed all others in both their incidence and importance in training. In conjunction with the other findings pertaining to the Massachusetts managers, it is evident that a large proportion of these managers had 'higher' qualifications than their Northern Ireland counterparts and this had important implications for the preponderance of R&D as well as general innovative activity within the firm. Massachusetts managers have a greater propensity to search for ways by which advances can be achieved, whether this is through product or process developments, or the acquisition of 'knowledge' which may be beneficial to the strategic direction of the company.

Storey (1994:149) has pointed out that in determining corporate strategy, the incidence of 'innovation and the identification of a particular niche' are key strategies associated with faster growth within small firms. Therefore the skills acquired through training may be attributed a crucial role in explaining the disparity in growth between the Northern Ireland and Massachusetts firms.

This is manifest in the Massachusetts managers attending a higher proportion of courses/seminars/workshops/lectures etc, which are off-the-job than the Northern Ireland managers, and by doing so, they benefit from the external economies of information flows which occur between firms operating within similar market structures. Conversely, Northern Ireland firms, by not exploiting the off-the-job training element renounce the potential benefits which can accrue from this activity through 'networking' (Mintzberg 1975).

Managerial competence on the shopfloor

The percentage of shop floor tasks which the managers could perform, provides an estimate of the degree to which managers were functionally flexible. It also indicates the extent to which white-collar workers were performing (or capable of performing) production functions. The post-Fordist theorists would propose that managers, that is the core workers (Atkinson, 1985), would be functionally flexible throughout the firm and therefore not only capable of planning, organizing and coordinating work, but also be competent in the technical aspects of production. Managers within the flexible firm should be 'team resource managers' (Bramlette, 1984), capable of working with shopfloor teams and providing diagnostic skills in the identification of inefficiencies (Papadimitriou, 1986). Conversely, within firms which epitomize Fordist forms of production, managers will be typically overseers and planners, with limited aptitude for shopfloor skills.

The managing directors of each of the matched plants identified the proportion of shopfloor skills which each of the managers could perform. Within the engineering matched plants, the average manager within Northern Ireland was capable of 64.6 per cent of shopfloor tasks, while the comparative figure for Massachusetts was 84 per cent of tasks.

This result, while being highly subjective, would suggest that Massachusetts managers were more adaptable throughout the firm than Northern Ireland managers. Within the Massachusetts firms the hierarchy of management had been restructured to a greater extent than within the Northern Ireland firms and consequently a higher proportion of Massachusetts managers were involved with shopfloor activities. Through their periodic involvement on the shopfloor they learnt and improved their production skills across a wide range of shopfloor tasks.

The ability of Massachusetts engineering managers to perform a greater proportion of shopfloor tasks is consistent with the results reflecting higher qualification levels amongst Massachusetts managers; the greater percentage of training which is done off-the-job; the larger quantity of money invested in the training of managers and the nature of the skills which managers acquired through training.

The greater ability of the Massachusetts engineering managers to perform shopfloor tasks is evident in 86 per cent of managers within Massachusetts being actively involved in the training of supervisors and shopfloor workers. This compares to 70 per cent of managers within the Northern Ireland engineering firms. Although the differential between the two regions is small within the engineering sector, it is significant in that again, it reflects a higher ability level amongst managers within Massachusetts to train other workers. The ability to train other workers is a direct function of the capacity of managers to perform the tasks which these workers are doing. Therefore, as managers within Massachusetts could perform a higher proportion of shopfloor tasks than their Northern Ireland counterparts, it follows that a higher proportion of managers within Massachusetts will be *capable* of training other workers[11].

Amongst the food processing firms, the results show that the average Northern Ireland manager could perform 79.6 per cent of the shopfloor tasks, as compared to 93.5 per cent by their Massachusetts counterparts. Further, within the Northern Ireland food processing firms, 86.6 per cent of the sample, or 13 of the 15 firms interviewed, stated that their managers took part in the training of other workers, while the comparative figure for Massachusetts managers was 93.3 per cent, or 14 of the 15 firms visited.

Acknowledging the subjectivity of these results, it is evident that the Massachusetts managers within the food processing firms were more flexible throughout the firm, as measured by their ability to carry-out shopfloor tasks, yet there is little difference between the regions in the proportion of managers who train shopfloor workers.

Summary

Atkinson (1985:20) has described the 'core' group as consisting of; 'full-time permanent career employees, say managers... Their employment security is won at the cost of accepting functional flexibility...[and]...the central characteristic of this group is their skills cannot readily be bought-in'. For managers to be characterized as the 'core' group of workers, it is implicit therefore, that they are functionally flexible throughout the firm. Subsequently, to attain this flexibility, they will have to display a wide range of skills with vertical integration being characteristic of their work practices.[12]

The findings from the matched plants have demonstrated a disparity in the presence of a core group of workers; the degree of functional flexibility; and the skill levels of managers, between the matched plants. Unequivocably, the findings illustrate that the Massachusetts managers approached each of these ideals to a much greater extent that their Northern Ireland counterparts.

Within small firms, limited time and money has often prevented managers from taking advantage of training courses. The benefits from training, in terms of corporate competitiveness are therefore forfeited (OTA, 1990). Indeed, where additional training is undertaken this may be designed to solve short-term problems, as opposed to the development of skills which will assist in the overall advancement of the firm. For example, the majority of training for managers within both the engineering and food processing firms in Northern Ireland was targeted at the development of technical skills whereas the nurturing of process development skills and a greater appreciation of the industry were the main skills which Massachusetts managers acquired from training. Within the Massachusetts matched plants, not only was there a higher per capita investment in managerial training, but this training was broader in scope to that undertaken by the Northern Ireland managers. The Massachusetts managers within the engineering firms (and to a lesser extent in the food processing firms) acquired cognitive skills which allowed them to evaluate every stage of the business, from production to white-collar tasks, and identify where changes could be made. Further, knowledge based training enabled them to be more aware of their market environment and target their resources more efficiently[13].

Bennett (1993) highlights a study carried out by MINTEL (1990) which illustrates that 72 per cent of managers within the UK felt that they were qualified enough in terms of their present jobs, with little apparent desire to retrain. The survey also showed that managers were more 'complacent' than any other group of workers about their skill levels and allowing for the fact that seven out of 10 managers by the year 2000 were already within the labour force in 1987 this would indicate that within these firms, the development of a core workforce is unlikely in the short-term. Similarly this matched plant study found that Northern Ireland managers were much more complacent about their skill levels than the Massachusetts managers, who instead, perceived managerial training as a way of nurturing their skills to the overall advancement of the firms, as opposed to the elimination of immediate problems.

The DTI (as quoted in O'Neill and MacNabb, 1994:14) have pointed out that the level of managerial training in Northern Ireland is minimal, and it suggests that anywhere between one in ten, to one in three business failures are as a direct result of inadequate training. Additionally, O'Neill and MacNabb point out that research by NISBI/NIERC in Northern Ireland saw lack of money and time, along with managers being unable to commission other workers to be 'in charge' of the business, cost of training as well as ignorance of the available training schemes - as the primary causes of the low uptake of training courses. However, this matched plant analysis has demonstrated that Massachusetts firms, operating within the same industrial sectors and displaying similar characteristics as the Northern Ireland firms, had overcome these constraints and were investing significantly more in the training of their managers.

To summarize, within the context of the matched plants, the Massachusetts engineering and food processing managers were functionally more flexible throughout the whole firm, being capable of a range of tasks throughout the firm and across occupational boundaries This has important implications for the reduction of both errors and subsequent down-time within their firms, as managers, acting as team resource people (Bramlette, 1994), not only plan, design and organize and evaluate production, but are also skilled to work with teams for problem solving and trouble shooting whenever needed. The Northern Ireland managers within both the engineering and food processing firms undertook less training and could perform fewer shopfloor tasks than their Massachusetts counterparts. A greater proportion of managerial training within the Northern Ireland firms was performed on-the-job, with technical skills being the most significant skills group. The lower ability of Northern Ireland managers to perform functions throughout the shopfloor therefore suggests that managers were performing traditional functions of planning, design coordination and control (Fayol, 1916), as opposed to a functionally flexible role whereby the demarcation of conception and execution was blurred.

Notes

1 Such employees did receive on-the-job training which provided them with the 'specific' skills needed to perform their jobs.

2 The factors which are highlighted represent brief summaries of results found within the research of Northern Ireland and Massachusetts. However, as the labour process is not an 'island' disconnected from other factors impinging upon the operations of the firm, it is essential to acknowledge the inter-relatedness of the variables and within this context how other factors may determine the regional perception towards training.

3 Curran et al. (1993) also found that employees were very reluctant to undergo training and this was equally as important as the unwillingness of employers to provide training.

4 Where reference is made to Massachusetts expenditure as defined by £ Sterling, then an average exchange rate of currencies is used of £1: $1:50 - the average value throughout the period of most of the interviews i.e. Spring to Autumn 1993.

5 These results are similar to the US pattern where it was found that two-thirds of all training dollars expended by businesses went to college-educated employees, who represented only 30 per cemt of the workforce and less than 10 per cent of all business training dollars went to blue-collar, entry level workers. (Carson; Journal of Educational Research 1993:296).

6 While it would be plausible to suggest that the smaller incidence of breakdowns within Massachusetts firms is due to the presence of more modern machinery within Massachusetts, this is not found in reality, but rather Northern Ireland firms slightly exceeded the Massachusetts firms in the percentage of machinery under 5 years of age. Therefore the disparity in the incidence of down-time as a result of technological failures is attributable to factors such as those outlined above.

7 Within large enterprises firms are more likely to advertise for suitably qualified applicants to become supervisors. In contrast, small firms restricted in terms of the time which is necessary in such a selection process, as well as wanting to avoid the costs of training 'new recruits', will tend to promote from within the business. Through 'promoting from within'

managers are also more aware of the other employee characteristics such as loyalty to the firm, conscientiousness, good worker relations, innovativeness and ability to foresee and overcome problems.

8 Caution should be exercized in the interpretation of this, in light of the findings of other variables indicating that the Massachusetts sample had a lower reject rate of production, and that Massachusetts firms operated a system of 'quality circles' through the interdependent relationship of shopfloor workers (see Chapter Seven). Additionally supervisors within Massachusetts were responsible for a larger number of shopfloor workers than that found within Northern Ireland firms, a factor which reflects the more prominent role of quality control amongst shopfloor workers as opposed to the Northern Ireland alternative which leaves the monitoring of quality within the remit of the supervisors.

9 Due to the firms being matched by employment size a similar proportion of managers also implies a similar number of managers within the Northern Ireland and Massachusetts firms.

10 It should be remembered that when new machinery is purchased, it is generally accompanied by training for employees as part of the capital investment.

11 The extent to which managers are involved in training workers is quite high within both the Northern Ireland and Massachusetts engineering firms. This is however a reflection of the size of the firms in that small firms, constrained by limited resources will depend more upon existing skills, than external consultants or training providers (Blackburn, 1990).

12 Vertical integration occurs when the gap between the conception and execution of ideas narrows. Therefore within the context of the managers, they will be capable of performing a range of functions, from planning and design, to the fulfilling of shopfloor tasks.

13 Other studies have also identified that small firms are often unaware or misguided in their perception of the competitors (Hewitt-Dundas et al., 1997) and of the demands of their customers (Roper et al., 1997). These studies also concluded that where such information gaps existed this would have an effect on the future competitiveness of the firms.

6 Numerical flexibility in the production process

Numerical flexibility

A working definition

Increasing instability in the relationship between markets, labour and production has created an uncertain and variable market environment within which firms must organize their production in the most efficient manner. Flexibility on the extensive margin (Hart, 1987) has been identified as a mechanism by which firms may adjust their labour requirements in response to changes in demand. This form of flexibility within the labour process has also been referred to as external flexibility (Streeck, 1987:290) or numerical flexibility (Atkinson, 1985, 1988; NEDO, 1986), and is seen as crucial to the ability of firms to overcome the crisis tendencies of Fordism. For example, firms may use numerical flexibility to respond 'more quickly, smoothly and cheaply' (National Economic Development Office, 1986:83) to changing market demands and so align labour costs and production costs.

Numerical flexibility can be defined as, 'The ability of firms to adjust their labour inputs over time to meet fluctuations in output e.g. overtime, flexi-time, new shift patterns, part-time workers, temporary workers, casual workers and subcontracting to other businesses' (Pinch, Mason and Witt, 1991:208).

For firms implementing such flexibility, numerical flexibility has been identified as being 'central to the strategy of overcoming crisis tendencies' within the capitalist system (Moulaert and Swyngedouw, 1989:338). The Regulation theorists also acknowledge that the adoption of 'flexible' policies will generate contradictions within capitalism, in that workers will resist the implementation of numerically flexible practices as they create uncertainty with respect to both income and employment. Therefore, it is likely that as within the Fordist era when the introduction of Keynesian economic management provided relatively stable product markets for Fordist firms, within the post-Fordist period a 'mode of social regulation' will evolve which through the provision of a surplus of unemployed people and the corporate restructuring of larger firms will reduce the capacity of workers to resist the introduction of flexible labour practices. Jessop (1990:309) states that the modes of regulation 'reinforce and underpin these regimes [of accumulation] by institutionalizing class struggle and confining it within certain parameters compatible with continuing accumulation'.

Unemployment, both cyclical and frictional, may increase and workers on the periphery of the labour force may experience considerable economic insecurity with periods of unemployment being punctuated with periods of employment (Gaffikin and Morrissey, 1992:27). The creation of an environment which is conducive to flexible labour practices is therefore characterized by increasing levels of unemployment; legislation which facilitates the use of flexible labour, and the decentralization of collective bargaining.

Within both the United Kingdom and the United States, there has been an increase in the levels of unemployed workers over the past 20 years. By 1991 the unemployment rate in Northern Ireland was 15.7 per cent (UK 7.1 per cent), compared to 9.0 per cent in Massachusetts (6.7 per cent in US). In Northern Ireland the proportion of long-term unemployment relative to total unemployment was also higher than in the remainder of the UK. Both factors suggest that within the workforce of Northern Ireland the basis has been laid for the implementation of numerically flexible practices.

In line with the belief that numerical flexibility will be most prevalent where there is a high incidence of unemployment along with a large distribution of small firms[1], it would be expected that there would be a relatively high incidence of numerical flexibility within the small firm sector of both Northern Ireland and Massachusetts.

This chapter builds on the findings of chapter five where the Massachusetts matched plants displayed significantly higher levels of labour flexibility, as measured through functional flexibility, than the Northern Ireland matched plants. This chapter examines the extent to which numerical flexibility was found within the matched plants and if the higher level of functional flexibility within the Massachusetts plants is replicated for numerical flexibility.

Previous international examinations of flexibility have been criticized for ignoring differences in the underlying institutional structures which impinge upon flexibility. For example, Pollert (1991:7) criticizes the investigation of labour market flexibility by the OECD (1989) and states that, 'those believing they are sharing the same debate on flexibility at an international level may be describing quite different processes within incommensurable economic, legal and institutional contexts'. While a comprehensive examination of the underlying institutional structures within Northern Ireland and Massachusetts is not feasible here, account is made of the limitations within the matched plants of the ability of firms to introduce numerical flexibility as constrained by the presence of union participation and activity within the firms. For example, where a union was present, did this correspond to the restricted introduction of numerical flexibility? Further, as Metcalf (1988:222) has highlighted, 'union presence in the workplace or company is associated with lower productivity and higher pay relative to non-union firms. Consequently, financial performance and profitability are lower where unions are recognized for collective bargaining'. To what extent therefore does this assertion also hold true for the sample of small engineering and food processing firms in Northern Ireland and Massachusetts?

Numerical flexibility within the matched plants

Part-time employment

As was outlined in the definition of numerical flexibility, flexible labour is characterized by an increasing use of part-time as opposed to full-time employees. Numerically flexible labour practices enable employers to hire workers when production requirements are high, while simultaneously laying these workers off during less prosperous periods. Firms will therefore minimize their labour costs of production and reciprocally maximize profits through the elimination of x-efficiency[2].

Within the sample of firms, workers were classified as part-time if they worked less than 30 hours per week. Although the use of part-time workers had not been adopted extensively, throughout the engineering firms, there was a greater incidence of part-time workers within the Massachusett's firms, than the Northern Ireland firms. Of the 35 engineering matched plants, only 3 employees (0.3 per cent of engineering employees) were classified as part-time workers within Northern Ireland, with the matching figure within Massachusetts being 18 employees (1.7 per cent of the engineering employees).

Within the food processing industry a higher incidence of part-time workers was found. Of the 15 Matched Plants, there were 57 part-time workers within the Northern Ireland firms and 46 part-time workers within the Massachusetts firms. Part-time workers accounted for 12.1 per cent of all shop floor employees within the Northern Ireland food processing firms, while the 46 part-time workers within the Massachusetts firms accounted for 9.8 per cent of all shop floor employees. Further, if the percentage of part-time shop floor workers is expressed as a percentage of the total employment within the 15 firms then the average Northern Ireland firm employs 10 per cent of its total workforce as part-time shopfloor employees, with the value for Massachusetts being 8.0 per cent.

The incidence of part time working reflects sectoral characteristics. Food processing is low skilled and through the utilization of standard machinery demands a low level of education or training from its employees. Consequently, it is easy for firms to 'hire and fire' as a direct response to changing market conditions. Conversely, within engineering there is a "low proportion of unskilled tasks, with most employees working with skill-specific technology. A high level of aptitude is therefore demanded from engineering employees with this aptitude being attained through training and experience. Within the engineering industry therefore, it would be much more difficult to operate a policy of numerical flexibility as this would directly affect both quality and productivity.[3] Data from the matched plants would therefore suggest that there will be a greater propensity for firms to use part-time workers where tasks are relatively low-skilled, training requirements for new recruits are minimal and there is a ready supply of available workers within the external labour market.

Whereas the use of part-time labour has until recently been found within the service sector, its acclaimed introduction and increased incidence within the manufacturing sector, must be viewed with caution. As the results for the small engineering firms within the survey illustrate, the use of part-time and numerically flexible labour within this industrial grouping is minimal and its lack of use is justified on the basis that it reduces the overall efficiency of the firm, as workers must be trained in the practices which they are involved in.

Temporary employment

Temporary employment practices may be used by firms where demand is uncertain or variable. Temporary workers are those employees who are recruited by a firm for a fixed period of time, after which employment may either be terminated, or extended.[4] The use of temporary contracts enables firms to combine their labour and capital resources more efficiently and determine the ability of employees to perform their jobs prior to them being granted permanent contracts. In general, where demand is uncertain, temporary workers will be characterized by fixed-term contracts. Where demand is uncertain temporary employment will be manifest through the greater use of seasonal workers.

It would be anticipated that firms pursuing a policy of corporate flexibility would have a significant proportion of temporary employees relative to their total work forces. Demand uncertainty amongst engineering firms and demand variability amongst food processing firms may prompt the increased use of fixed-term and seasonal contracts respectively. Through temporary contracts, capacity utilization can be coupled more closely with the utilization of labour, allowing a quick and efficient response to changing market demands.

Within the sample of engineering firms in Northern Ireland, 0.4 per cent of the workforce were temporary workers with the corresponding figure for the

Massachusetts engineering firms being 2.8 per cent. In contrast to the engineering sector, 6.1 per cent of the workers within the food processing sector in the Northern Ireland sample were temporary workers, with 5.7 per cent of the Massachusetts workers being temporary workers.

As for part-time workers, again a sectoral disparity was found in the presence of temporary workers within the matched plants. As was anticipated, the food processing sector had a much higher incidence of temporary employees than the engineering sector with this being common to both Northern Ireland and Massachusetts firms. While there was a slightly higher incidence of temporary workers within the Massachusetts engineering firms than the Northern Ireland engineering firms, this was not statistically significant.

This sectoral variation can be explained in terms of the skill levels required by 'peripheral', or temporary workers. For example, within the food processing sector, the majority of shopfloor employees are unskilled and perform tasks which through the use of dedicated machinery, are highly repetitive and low-skilled. As many of the interviewees indicated, food processing firms hire people who have no formal qualifications. Through the use of temporary contracts for 'new-starts', the suitability of employees can therefore be evaluated, both in terms of their accuracy and speed in performing rote tasks. Those employees which prove unsuitable can then be released at the end of their contracts without the mandatory regulations for the termination of employment.

Within the engineering sector, there is a lower proportion of unskilled and low-skilled jobs and therefore employers demand proof of a minimum level of aptitude from their workers prior to employment. As aptitudes can be assessed in terms of vocational qualifications or prior work experience, it is easier for employers to determine the suitability of candidates for employment prior to recruitment. Where temporary contracts are used as a means of assessing competence, sectoral variations are therefore expected. Further, as for the incidence of part-time workers, in the absence of a suitably skilled external labour supply, the incidence of temporary workers will be constrained by the skill barriers to entry.

The findings from the matched plants suggest that demand uncertainty and variability were of only slight importance in explaining the incidence of temporary workers within the sample of firms. Instead, temporary contracts were used within the food processing firms as a measure of employee competence. Within the engineering firms, the ability of firms to determine the competences of employees prior to recruitment, combined with a 'skill barrier to entry' for labour, reduced their propensity to use temporary labour.

The ability of firms to introduce flexible working practices will be strongly influenced by the socio-economic environment. Where there are high levels of unemployment the bargaining position of unions will be considerably weakened. In addition, fewer opportunities for employees to formulate collective agreements within small firms will result in a lower incidence of unions within small as compared to large firms (Daniel and Millward, 1983). As both Northern Ireland and Massachusetts have experienced disproportionately high levels of unemployment and as the investigation is based within the context of the small firm sector, it would be anticipated that the presence of collective power, in the form of unions, would be low within both Northern Ireland and Massachusetts. Where unionization is low, there will be a higher potential for the introduction of flexible working practices.

Twenty-three per cent of Northern Ireland engineering firms had a union representation amongst their employees, while the corresponding figure for Massachusetts was 43 per cent. Examination of the engineering firms by sector demonstrates that union activity within the metal products firms was similar for both regions, with 25 per cent of the firms having a union present. Within the mechanical engineering and electrical and electronics sectors, differences were evident between the regions. For both of these sectors, 22 per cent of Northern Ireland firms, as compared to 55 per cent within Massachusetts had a union representation within their plants.

The food processing sector also displayed regional differences in the extent of union activity. Within Northern Ireland small food processing firms, only one of the 15 firms visited, had workers which belonged to a union. Conversely, within Massachusetts 14 of the 15 firms visited had a union representation within the firm.

While traditionally, union membership has been limited amongst employees within small firms (Daniel and Millward, 1983), the differences between the regions in terms of union activity cannot be related to the size of the firm, as firms were matched by employment levels. An explanation of this anomaly may be found within the political environments of the matched plants. For example, Katz (1993) found that within both the United Kingdom and the United States of America there has been a significant move away from collective bargaining since 1980.

Farnham (1990) proposes that the incidence of collective power would reduce the incidence of numerical flexibility within the firm, however this is not supported within the survey findings. Instead, in terms of both part-time workers and temporary workers, the engineering firms in Massachusetts exceeded the corresponding figures within Northern Ireland.[5] It should be noted that this does not necessarily invalidate the general proposition made by Farnham, but instead casts some doubt on its applicability within the small firm sector. In terms of the food processing sector, a corresponding proportion of both part-time and temporary workers was found for both regions, and this would suggest that, although a major difference existed in the incidence of a 'union' within the matched plants, this had no effect on the nature of employment adopted. A clear relationship could not be identified therefore between union activity and the implementation of numerical flexibility.

Regional and sectoral disparities in pay

Within Atkinson's model of the flexible firm, there is a distinction between the core and peripheral workforce. Core workers are associated with high skills; training; employment security and attractive levels of remuneration. In contrast, peripheral workers are associated with low-skills and little or no training; employment insecurity; and a disproportionately low level of remuneration relative to the core. Should the sample of firms conform to the pattern of the flexible firm, then it would be anticipated that evidence of a polarization in wages would be apparent between the skilled and unskilled workers. Further, it would be expected that where unions were active within the firms there would be a smaller gap in pay between skilled and unskilled workers and overall levels of remuneration would be higher (Metcalf, 1988).

The rates of pay of employees within engineering and food processing firms were examined, and the mean hourly rates for unskilled; semi-skilled; skilled employees and technicians; and managerial workers attained (Table 6.1). It should be noted that where yearly salary statistics were provided by the interviewees, then the hourly rate was estimated on the basis of a 40 hour week, for 52 weeks per year.

Table 6.1
The average hourly pay levels of employees (1993)

	Northern Ireland	Massachu-setts	Northern Ireland	Massachu-setts
	Engineering		*Food Processing*	
Unskilled	4.05	3.72	2.87	4.00
Semi-skilled	5.93	6.80	4.75	5.33
Skilled	7.63	10.03	5.25	7.66
Managerial	10.45	13.70	6.5	11.21

Notes: Pay levels are in £ Sterling where exchange rate is based upon £1:$1:50.

Source: Matched plant survey

With the exception of unskilled workers within the engineering industry, Massachusetts employees earned a higher wage than their Northern Ireland counterparts within both the engineering and food processing sectors. Further, the differential between the wages for Northern Ireland and Massachusetts employees increased in line with the increase in skill levels. For example, within the small engineering firms, unskilled employees within Massachusetts earned 92 per cent that of the wage levels of Northern Ireland unskilled employees; semi-skilled Massachusetts workers earned 114 per cent that of their Northern Ireland counterparts; and both skilled and managerial workers within Massachusetts received wage levels equivalent to 131 per cent that of the Northern Ireland skilled workers and managers.

Within the food processing sector, a similar pattern was found with unskilled Massachusetts workers earning 139 per cent that of Northern Ireland unskilled workers; semi-skilled workers earning 112 per cent that of Northern Ireland semi-skilled workers; skilled workers receiving 146 per cent that of skilled workers within Northern Ireland; and managerial workers within Massachusetts earning 172 per cent that of managerial workers within Northern Ireland.

In absolute terms it is evident that the monetary value of wage rates within Massachusetts were higher than that received within Northern Ireland[6]. Of greater importance in determining whether or not there was a polarization in wages between the core and the unskilled peripheral workers, the gradient between the pay levels of unskilled and other skill groups was calculated (Table 6.2).

Table 6.2
Pay levels of semi-skilled, skilled and managerial workers relative to unskilled workers

	Northern Ireland	Massachu-setts	Northern Ireland	Massachu-setts
	Engineering		*Food Processing*	
Unskilled = 1.00				
Semi-skilled	1.46	1.85	1.66	1.33
Skilled	1.88	2.70	1.83	1.91
Managerial	2.58	3.70	2.27	2.81

Source: Matched plant survey

The greater the value in Table 6.2, the higher the differential between the pay levels of unskilled and other employees. For the engineering sector there was a smaller differential between the pay levels of unskilled and other workers within the Northern Ireland firms as compared to the Massachusetts firms. For example, within the engineering industry, unskilled workers within Northern Ireland earned 68.3 per cent of the pay levels of semi-skilled workers, 53 per cent of skilled workers and 38.7 per cent of managers. Conversely, within Massachusetts, unskilled employees within the engineering sector received a lower proportion of remuneration, relative to all the other skill groups, than their Northern Ireland counterparts. For example, within Massachusetts unskilled employees within the engineering industry earned 54 per cent that of the semi-skilled employees; 37 per cent that of skilled employees and only 27 per cent that of managerial workers. Therefore in comparative terms, there was a much larger differential between the pay levels of Massachusetts engineering employees than that of Northern Ireland engineering employees.

The results also show that amongst the food processing firms again there was a larger differential in pay between unskilled and other skill groups, within the Massachusetts firms. For example, unskilled workers within Massachusetts earned 52.2 per cent that of skilled workers, and 35.6 per cent that of managerial workers, while the corresponding figures for Northern Ireland were

54.7 per cent and 44 per cent respectively. Therefore, while the differential in pay levels of unskilled employees as compared to other skills groups for the food processing sector, was not as large as that found within the engineering sector, it was still large enough to suggest that there is a peripheralization of unskilled workers relative to other employees. While it has been illustrated that unskilled workers within Massachusetts earn a wage which represents a greater differential between this group and all other skill groups, the differential between unskilled and semi-skilled workers within the food processing sector of Massachusetts was smaller than that of Northern Ireland. That is, the food processing unskilled workers within Massachusetts earned 75 per cent of the semi-skilled wage, whereas within Northern Ireland this same skill group earned only 60 per cent of the semi-skilled wage. It is more accurate to perceive the peripheralized group within the Massachusetts food processing sector as consisting of both unskilled and semi-skilled workers, as opposed to just unskilled workers within the Massachusetts engineering industry.

Results from the sample failed to support the hypothesis that a higher incidence of collective power would manifest itself in higher wages (Metcalf, 1988:222). In general, where unions are in operation, they will seek to maximize the benefits, and optimize the conditions of all employees, irrespective of skill levels. For example, in the United Kingdom the Amalgamated Engineering Union (AEU) in its company guidelines for the attainment of Total Productive Maintenance unequivocably includes *all* employees within the restructuring process, and seeks to serve equally each of its members in the restructuring of the firm. However, within the firms in Massachusetts, although there was a high 'collective' presence as manifest through the extent of union membership, this was not translated into higher wages, and instead a greater peripheralization of unskilled (in the case of the engineering sector) and unskilled and semi-skilled workers (in the case of the food processing sector) had resulted.

Summary

If flexibility is gauged by the model of the 'flexible firm' (Atkinson, 1984), it can be concluded that while some characteristics of the matched plants conform to the model, the extent of conformance is somewhat limited. Although no distinction is made between the applicability of the model within different sectors the matched plant findings demonstrate significant differences in the incidence of the measures of flexibility.

For each of the four indicators of numerical flexibility namely, the incidence of part-time workers; the incidence of temporary workers; the collectivization of power; and disparities in pay, significant differences were apparent between the engineering and food processing firms. Food processing firms within both

Northern Ireland and Massachusetts had a higher proportion of their work forces employed as part-time or temporary workers. While the flexible firm model suggests that firms will use an increasing number of part-time and temporary workers in response to demand uncertainty and variability, this waas not found to be a decisive issue in the employment of such workers.

Where skill barriers to entry existed, such as amongst the engineering firms, this constrained the ability of employers to recruit in response to changing demands. Where firms had to make investments in the induction and training of new recruits, this also reduced the extent to which numerically flexible labour was used.

A number of the food processing firms acknowledged that temporary contracts enabled them to assess the suitability and aptitude of employees prior to being hired on a permanent basis. Demand variability was not the most important issue in prompting the use of such labour. Testing the suitability of employees was not as significant an issue within the engineering firms as recruits within this sector were required to demonstrate their aptitudes (either through vocational qualifications or experience) prior to recruitment.

Findings from the matched plants also demonstrated that the incidence of collective power within the firms did not lead to greater equality of pay amongst skill groups. Amongst the Massachusetts firms a high level of union activity was associated with greater disparities in pay. Pay disparities within both the Massachusetts engineering and food processing firms suggest that a peripheral and core workforce may exist. The other measures of numerical flexibility however, provide little support for this.

Numerical flexibility is however only one aspect of the 'flexible firm' model. There are dangers in the analysis of labour flexibility that aspects of the same equation may be studied disproportionately, and assumptions thus made which are unrepresentative of the whole situation. Numerical flexibility must therefore be examined as an integral part of labour flexibility in conjunction with functional flexibility. Numerical flexibility and functional flexibility are not mutually exclusive, but rather, as other commentators have found (Guidetti, 1995), are intricately interwoven into each other.

Labour flexibility: Conclusions

Flexibility at the micro-level is measured by the ability of the firm to respond and adapt to changes within the market. Uncertainty of demand makes it essential for firms to eliminate rigidities within the production process. The Flexible Specializationists perceive the introduction of flexible technology as demanding changes in the organization of labour within the firm (Piore and Sabel, 1984: Sayer, 1985), while the Regulationists although attributing a lesser role to technology in stimulating a crisis of Fordism, also acknowledge that labour flexibility will increase with the traditional division of labour being reversed through job enrichment and multi-skilling (Boyer, 1988; Palloix, 1976). Similarly, both the Flexible Specializationists and Regulationists see the restructuring of labour as involving the organization of work around semi-autonomous or flexible work groups, or teams (Aglietta, 1979; Blackburn et al., 1985; Murray, 1988; 1989).

Despite the Flexible Specializationists and Regulationists proposing that changes are likely in the organization of labour, neither approach offers a concise summary of these changes. For this reason the flexible firm model (Atkinson, 1984; 1985) has been used extensively as an 'analytical tool' (Atkinson and Meager, NEDO, 1986) by which labour flexibility can be gauged.

The ability of the flexible firm to adjust to market uncertainty is achieved through the use of numerically and functionally flexible labour. Numerical flexibility is sought from peripheral workers 'so that headcount can be quickly and easily increased or decreased in line with even short-term changes in the level of demand for labour' (Atkinson, 1984). Conversely, within functional flexibility, the core workers which are central to the firms activities are provided with job security in return for their crossing of occupational boundaries which enables the firm to adjust more closely to production demands. The principles embodied within the flexible firm model were therefore used in the matched plant examination of labour flexibility within small engineering and food processing firms in Northern Ireland and Massachusetts.

The Flexible Specializationists see flexible technology as enabling small firms in particular, to gain competitive advantages based upon economies of scope, as compared to the traditional over-dependence upon economies of scale. It would be anticipated therefore that small firms, by virtue of their size would not have managerial structures and lines of job demarcation to the same extent as within large firms. These firms would therefore be the most likely to introduce numerical and functional flexibility.

The matched plant analysis of small engineering and food processing firms in Northern Ireland and Massachusetts did not, however, display significant evidence for numerical flexibility. While the flexible firm model proposes that uncertainty or variability in market demand will lead to firms developing a peripheral workforce, composed of part-time and temporary workers, this was not evident within the matched plants. Skill barriers to entry; the extent to which new recruits needed training and the availability of other suitably trained workers in the open market were the major determinants of numerical flexibility. These factors created sectoral differences in the sample, with the food processing firms displaying greater numerical flexibility.

Analysis of disparities in pay, did however suggest the polarization of unskilled relative to highly-skilled workers. The marginalization of unskilled workers was most apparent within the Massachusetts firms, being found within both the engineering and food processing samples. Further evidence of marginalization of Massachusetts unskilled workers was not found, but instead these workers displayed greater functional flexibility than their Northern Ireland counterparts, as measured by; the elimination of job demarcations; the extent to which employees interacted; educational levels and investments in training.

Evidence from the matched plants also demonstrated that job-loadings, as characterized by the vertical integration of tasks were not occurring within firms in Northern Ireland, but were evident, within both engineering and, (to a lesser extent) food processing firms in Massachusetts. Also, the training of supervisory workers within Massachusetts was much more advanced than that found within Northern Ireland, in terms of the per capita expenditure on training; qualification levels; the utilization of off-the-job training (which is indicative of the movement towards functional flexibility - Foster, 1988); the skills which were demanded of supervisors; and the ability of supervisors to cross occupational boundaries within the firm. However, simultaneously, supervisors within Massachusetts had experienced widespread corporate restructuring, with many supervisors losing their positions within the 'middle-tier' of the managerial hierarchy.

The flexible firm model would see these workers as being demoted to the periphery of the workforce, with de-skilling and employment insecurity being found. In reality however this was not found and instead, through training the skills of the Massachusetts supervisors were both enhanced and increased, as evident in their ability to perform cross trade/occupational jobs within the firms. Conversely within Northern Ireland, the supervisors, although provided with this middle managerial status in title, were merely acting in a traditional Fordist way, as leadmen or one-to-one supervisors (Bramlette, 1984).

Within the Northern Ireland matched plants, although there was a growing awareness amongst employers of the importance of training their work forces, the implementation of appropriate measures was restricted by the attitudes of the workers, as they refused to attend courses or further their skill levels. As a method to overcome this, the CEDEFOP (1990:34) have proposed that a fundamental restructuring of priorities needs to occur within the education systems. They refer to the traditional or conventional form of education as consisting of a base of theoretical knowledge, built upon practical knowledge and topped with attitudes and skills. Conversely, they propose that the 'Reformed Triad' will consist of a base of attitudes and skills, with practical knowledge built upon this, and topped with theoretical knowledge. Certainly within the context of the Northern Ireland matched plants, the reformed triad would be most helpful in altering the ability levels of employees. Unless attitudes and skills are conducive to the advancement of the economy through the acquisition of practical and theoretical knowledge, then the attainment of functional flexibility is merely an elusive concept.

The Northern Ireland managers had not developed a 'core' workforce, relative to their counterparts within Massachusetts. The principle of greater managerial involvement in training outlined within the Management Charter Initiative[7] (MCI) appears to be both abstract and remote to the managers within the manufacturing sample in Northern Ireland. In contrast, the Massachusetts firms were making substantial investments in managerial training with this being targeted, not only at technical skills, but also the attainment of knowledge of the industrial sector and organizational management.

Although managerial training within the Northern Ireland firms was predominantly technical, the Northern Ireland managers were less functionally flexible throughout the firms. These managers performed the traditional functions of planning, organizing, coordinating and controlling (Fayol, 1916) and depended upon supervisors and shopfloor workers to undertake production. There was therefore a greater demarcation between conception and execution of tasks within the Northern Ireland firms. In line with the view that the demarcation of tasks between white-collar work and production will be blurred within the flexible firm (Jessop et al., 1988), this supports other findings to conclude that while there was limited evidence of numerical flexibility (with any evidence of numeral flexibility being sectorally specific, greater evidence of functional flexibility was found within the Massachusetts matched plants. Yet, functional flexibility within the Massachusetts firms was not confined to the core, or managerial workers, but instead higher levels of skill enhancement and the formation of interdependent working was evident throughout the workforce.

Notes

1 This is a situation which is accentuated when the nature of unemployment is long-term and entrenched, which thus prevents the agglomeration of workers and the formulation of collective agreements.

2 Lower quasi-fixed labour costs have also been identified by Abraham (1987) as a major incentive to firms in using part-time and temporary workers. The use of numerically flexible labour will minimize costs directly through wages, employer's national insurance contributions and sick pay etc., and indirectly through the more efficient capital to labour ratio, leading to higher levels of added value and profit per employee.

3 Due to the investment needed in training new recruits and the subsequent time lag between their recruitment and their creation of added value, it is preferable for firms which require a significant level of skill from their employees not to pursue a part-time employment policy.

4 Where employment is 'extended' this may be either in the form of a further temporary contract or a permanent contract.

5 It should be remembered, that although the disparity between the absolute number of part-time and temporary employees found within engineering firms in Northern Ireland and

Massachusetts was small, when converted into percentage values such appeared significant, and as such is worthy of note. Additionally because the disparity was found in favour of Massachusetts for both forms of numerical flexibility, this also reaffirms the findings that there is a greater incidence of numerical flexibility within Massachusetts than that of Northern Ireland.

6 Higher absolute levels of remuneration within the Massachusetts firms does not necessarily imply that employees within Massachusetts were remunerated at a higher level than workers within Northern Ireland. For such a conclusion to be made, cost of living statistics would have to be accounted for.

7 The Management Charter Initiative (MCI) (1987) was the culmination of an investigation by the National Forum For Management Education and Development (NFMED) into the relative inferiority of training within the UK. The MCI was an initiative developed with the intention of increasing the participation levels of managers throughout all spheres of economic life, and was closely aligned with the National Vocational Qualifications (NVQs).

7 Flexibility in the production process

Introduction

Four scenarios have been proposed to explain the decline of the Fordist era (Boyer, 1991). The first, concerned with the 'exhaustion of the previous technical regime', proposes that the benefits of economies of scale and the consequent 'giantism' which had resulted, led to a decline in the output to capital ratio through a gradual reduction in productivity. The second scenario is based upon the social and economic contradiction of the methods of production within the Fordist era. These contradictions evolved through the inherent principles of 'Scientific Management', whereby the nature of production demanded that middle-management monitored and assessed blue-collar workers, and thus labour which was potentially productive assumed an unproductive role, leading to an overall reduction in labour productivity. Additionally, as markets became increasingly fragmented with consumers demanding more varied and customized products a conflict emerged between unpredictable consumer demand and dedicated, rigid capital.

The third scenario was the emerging 'conflict between rigid technologies and uncertain macroeconomic prospects' (Boyer, 1991), whereby the typical macroeconomic policies adopted during this period within countries such as the UK and the USA, were focused upon *stop-go* strategies. Yet, the nature of Fordism demanded stable markets and consistent demand from consumers, but this could not be attained in conjunction with the unpredictable national economic cycles of boom and depression. With the diversification of tastes and fashions a greater demand for differentiated products was emerging and subsequently, technologies based upon the production of cheap, mass produced goods were replaced by flexible and adaptable technologies.

The final scenario as proposed by Boyer (1991) to explain the causes of the demise of Fordism is rooted in the 'break-down of the international order...as regards the loss of international co-operation...economic policies, exchange rate dynamics (and) monetary controls' (Boyer, 1991:111), of which the collapse of the Bretton Woods Agreement played a prominent role. Boyer suggests that this *international disorder* was the primary cause of the crisis of Fordism as technologies became inherently rigid due to their non-transferability between nations. Additionally, a conflict resulted between the regulatory controls at the national level and the simultaneous development of an international regime of accumulation, which led to the non-correspondence between the two and the subsequent break-down of Fordism.

Jessop (1991:7) highlights that if competitiveness is to be achieved by the traditionally Fordist nations then a *radical approach* is required in the transformation of production. He states that, 'growth efficiency involves choosing among available production techniques and products, and innovative efficiency is concerned with innovation in processes, products, markets and organization.' Therefore the long-standing American economic hegemony of the 20th century is now in decline as the rules of the capitalist game have changed, and the pawns on the board which are ill-positioned to adapt to these changes are in danger of being permanently removed, with the subsequent decline in the economic status of their players. Hence, Jessop (1991:10) refers to the transition in production as a competitive struggle, 'to switch quickly and smoothly among innovative products and processes with each new product offering better functional qualities and improved efficiency in production'.

Flexible technology in the form of programmable automation (PA), computer numerically controlled (CNC) machine tools and flexible manufacturing systems are regarded as facilitators of this 'competitive struggle' (Kelley, 1989). In the extreme case flexible production would include the manufacture of small batches and a broad range of product types (Piore and Sabel, 1984). The dominance of large firms would therefore be reduced through the attainment of competitive advantages based upon economies of scope, as opposed to merely scale.

The introduction of flexible technology necessitates organizational innovations (Hirschhorn, 1984; Piore and Sabel, 1984) including a 'radical devolution of existing bureaucratic structures and a concomitant rise in collaborative, skill-enhancing forms of work organization' (Kelley, 1989:235). Flexible technology within the firm will therefore overcome Boyer's third crisis scenario, in enabling small firms to be competitive within increasingly fragmented and variable markets. Boyer's first crisis scenario of diminishing productivity will also be overcome through the more efficient utilization of resources (which include the elimination of buffer stocks) and the increasing importance of economies of scope in conjunction with those of scale. Finally, the organizational innovations associated with the introduction of flexible technology (including increased employee involvement and multi-skilling) will lead to the managerial hierarchy being restructured and the internal division of labour reduced.

Where previously firms would have held a large proportion of their raw materials and finished goods in-house, the new regime of flexible accumulation is characterized by a very low level of goods held as inventory. The monitoring of quality is also radically altered within this new production process with continuous checking and subsequent reworking being found as opposed to the traditional ex-post testing of quality which characterized the Fordist regime. Thus, as a result of these new methods of quality control, reject rates are lower within the new system and defective products are virtually eliminated from the output distributed to buyers. Conversely, within the Fordist regime defective products were frequently unidentified and subsequently concealed within the buffer stocks held within the firm.

Where previously there was a considerable loss of production time due to elongated periods of time to set-up machines, deal with inventory problems, cope with build-ups of production and maintain faulty machines, this is minimized within the new regime in an attempt to encourage the maximum value of productive labour from the workers. This minimization of the amount of lost productive time is achieved in a number of ways and are later outlined with respect to the sample of small firms in Northern Ireland and Massachusetts.

This chapter examines the extent of flexibility in the production process for the sample of matched plants in Northern Ireland and Massachusetts. Having outlined within chapters Five and Six the extent and incidence of labour flexibility, it follows that an analysis should be made of flexibility within production. Gertler (1988:426) states that 'flexible machines work best with flexible labour'. Therefore, in light of the presence of labour flexibility within the firms, this chapter examines whether or not there is a correlation between labour flexibility and productive flexibility. The characteristics of production under flexible (Just-In-Time), as compared to Fordist (Just-In-Case) principles of production are outlined, and these provide the benchmark against which flexibility in the production processes of the matched plants is measured. Flexibility within the production process is measured through five indicators including, the extent to which the matched plants manufacture for stock; the nature of subcontracting relationships; the incidence of research and development (R&D) activity within the firms; the manner in which quality control is incorporated within production; and the composition of the capital stock.

Just-in-time and Just-in-case production

Although the introduction of new technologies has attracted widespread interest, considerably less attention has been paid to relationships between workers; workers and management; firms and their buyers and suppliers (Sayer and Walker, 1992:162). The Western industrialized countries have increasingly looked to Japan to understand 'why Japanese business enterprises are so effective in the marketplace' (Best, 1990:137). Best concludes that Japanese firms 'compete strategically on the basis of price, product quality, technological process, or product innovation ... the firm is a learning organization that is continuously creating new productive services by teamwork and experience, and ... inter-firm relations can be cooperative or market oriented' (Best, 1990:166).

Concepts such as Just-In-Time production (JIT) although emanating from Japan are not abstract from the ideas proposed by the Flexible Specializationist school. The Flexible Specializationists in drawing upon the evidence in the Third Italy (Sabel, 1982; 1989) emphasize the creation of 'flexible production complexes'. The formation of industrial districts is seen as a model 'of economic co-ordination and social integration for the new specialist production system... seen to foster harmonious equilibrium between competition and co-operation between firms' (Burrows et al., 1992:3; Swyngedouw, 1991). The agglomeration of activities is regarded as central to the expansion of economic activity in 'diminishing temporal-spatial uncertainty' (Swyngedouw, 1991:309; Scott, 1983). Therefore, it is within these complexes that the principles of JIT will be most effective (Gough, 1992) in allowing firms to respond more rapidly to fragmented and volatile markets. Just-In-Time production (JIT) is therefore central to the introduction of flexibility both within the production process and between firms. To fully appreciate the nature of JIT, the characteristics of Just-In-Case (JIC) production should also be outlined.

The characteristics of JIC bear important similarities to that of Fordist production, for it is centered around the fragmentation of tasks, with workers being utilized as appendages of machines. Employees are assigned specific tasks which through repetition, they increase their speed and efficiency. Additionally, the capital machinery which these workers use is job-specific, with it being virtually impossible to alter the function which it performs. Due to the rigidity of the capital, it may take an extended period of time to set-up the machinery and get it to full speed, a factor which has important implications for the form of production.

Within the JIC system, dedicated machinery results in long production runs. The system was reactive rather than proactive, with preventative measures dominating production. These preventative measures took the form of the accumulation of 'buffer stocks' of goods or components in terms of finished goods within the manufacturing process. However, prevention against unforeseen circumstances was also apparent within the supply side of the firm, in relation to suppliers of raw materials and other elements of production.

JIC demanded that a stock of raw materials or parts from other suppliers was available within the firm, and therefore should changes in demand occur, or a problem arise within the supplier plant, then the producer would have adequate materials in-house to maintain production, at least in the short-run. Although this last aspect of JIC was production related, it was also largely determined by the lack of geographical concentration or relative dispersion of industry. Sayer and Walker state (1992:168) that 'distant and non-interventionist relations with suppliers had disadvantages; large, infrequent deliveries implied heavy warehousing and related costs; permissive attitudes to suppliers' quality control increased costs; and arms-length relations failed to encourage harmonization between supplier and customer, for example, in the design of components.' (Sayer and Walker, 1992:168).

Within the JIC organization of production there was a high demand for unproductive, supervisory labour to monitor the work of shopfloor employees and coordinate activities throughout the firm. This led to a 'deep vertical hierarchy of supervisory and managerial' control (Sayer and Walker, 1992:166), as manifest in clear lines of demarcation and limited interaction between blue collar shop-floor employees.

Further, within JIC, quality control is performed as a separate departmental function within the production process, with errors being identified by spot checks and testing of sample products at the end of the production. Subsequently, this meant that again personnel were being employed within unproductive functions and resources were being wasted through the concealment of inferior products within buffer stocks.

JIC production created major advances for the production process of the late 19th century and resulted in increased levels of productivity. Additionally, in the creation of mass produced standardized products JIC served many of the needs of industry by ensuring that stocks were always adequate to cover productive needs, should short-run conditions change rapidly, or faults arise in some aspects of the technology. However, JIT provided the firm with the potential to increase productivity further and overcome the contradictions abiding within JIC.

The JIT system of production is 'a way of organizing the immediate manufacturing labour process and buyer-supplier relationships between firms, but it is normally surrounded and supported by a wider set of practices regarding skills, labor-management relations, and labor market conditions.' (Sayer and Walker, 1992:170). The over-riding ideology of JIT stands in stark contrast to that of JIC, for contrary to the principles of JIC which were formulated around preventative measures, JIT is based on the concept of performing functions only when necessary. This 'last minute' dogma has therefore an association with the nature of production, in that goods are not produced prior to the attainment of orders, but rather as orders arise, then technological functions are adapted to equate supply with demand. Further, within JIT the maintenance of buffer stocks are eliminated due to the cost burden which such impose upon the firm, and therefore relationships are established with suppliers which encourage frequent and rapid delivery of goods whenever required by the producer.

As a result of the above two principles of JIT production, the labour requirements are significantly altered both in terms of every-day productive tasks and with respect to the assessment of quality. Whereas previously, under JIC workers were skilled only at one particular task which they performed routinely, workers within the JIT system are encouraged to learn a range of functions throughout the shop-floor, thus facilitating an increasing level of interaction between workers. Where previously the production process was characterized by conveyor belts which brought the work to the workers, and concentrated on specific functions such as welding, fabrication, cutting, grinding and polishing etc. this is replaced by machinery which is grouped according to its order within the production line, therefore there is a 'clumping' of processes within JIT as opposed to the traditional system of production evolution. Therefore within the JIT system workers are encouraged to move throughout the shop-floor as they interchange from one function to another.

With respect to the monitoring of quality, the 'quality department' is eliminated and instead workers are trained to monitor the quality of their own work at each stage of the production process, and thus the early detection of errors can facilitate the reworking of materials, and subsequently reduce wastage of resources. Further, workers are also trained to maintain their machinery, and therefore through cleaning, greasing, oiling etc., of machinery, the incidence of expensive breakdowns is reduced.

While it would be plausible to suggest that the JIC system of production was characterized by 'rigidity' in terms of the job specifications for individual employees; the need for large buffer stocks; the need for a large pool of unproductive labour to monitor other workers and perform quality control functions etc. it also displayed some elements of flexibility. For example, should changes suddenly arise within the market in terms of increased demand for specific products, these firms, due to their maintenance of large buffer stocks could supply the market and reap the associated profits of this.

The JIT system also, while being characterized by 'flexibility' in terms of its ability to respond quickly to conditions of both demand and supply, along with its flexible forms of labour control and work, also contains aspects which are extremely rigid in their structure. For example, the need to establish reliable relationships with suppliers may lead to the development of 'ties' and 'loyalties' to certain firms which is contrary to the notion of flexibility of production. Further, this 'networking' may take an extended period of time to develop and cultivate, time which could be used in more productive tasks.

Finally, through the minimization of stocks and the use of numerical flexibility with respect to the labour force, it is essential that production planning is performed with precision and accuracy, to avoid situations where output does not equate with demand, or the non-utilization of technology which could be profitably used to produce products for which there is an immediate demand (see Sayer and Walker, 1992).

Distribution of suppliers and customers

The introduction of flexible production methods (Scott, 1988) have not only affected the internal functioning of firms, but also the temporal and spatial 'choreography of production logistics' (Swyngedouw, 1991:326). The 'flexible production complex' as proposed by the Flexible Specialization school (Sabel, 1982; Brusco and Sabel, 1983; Piore and Sabel, 1984) in conjunction with the Japanese kanban, or Just-In-Time models of intra-firm and inter-firm organization, now challenge the traditional models of industrial location as based upon the product life cycle, the new international division of labour and the urban hierarchy models. Within flexible production it is the reconcentration, or integration, of production into regional complexes which determines their success (Scott, 1983; Storper and Scott, 1988). Storper and Walker (1989:152) outline that all flexible production industries are characterized by 'organizational fragmentation in which dense, unstandardized, transactional relations between firms are particularly important'. Due to both demand uncertainty and variability, territorial complexes provide a means of 'handling the numerous problems of interfirm organization of production' (Walker, 1989:87).

It would be expected therefore that where firms were implementing flexible production methods, there would also be close spatial proximity between customers and suppliers, the spatial integration of firms being necessary for firms to respond quickly to both product and process innovations. Yet, it is also possible to infer from this rationale that where industrial complexes do not exist, this will constrain the extent to which flexible production methods are utilized. For example, the Japanese form of JIT represents a method of integrating work between factories (Cusumano, 1985; Schonberger, 1982) and therefore where inter-firm relationships are curtailed due to the territorial organization of production it will be increasingly difficult to implement these principles. Indeed, the flexible specializationists argue that the development of inter-firm relations and the increasingly 'complex divisions of labour' (Walker, 1989:88) mean that in the absence of flexible production complexes, flexible production methods will not be found. It would therefore be anticipated that if the matched plants were located close to their suppliers and customers then this would be mirrored in the incidence of flexible production methods. The location of suppliers and customers to the matched plants was therefore examined.

Each of the firms identified the percentage of sales which was sold within four broad geographical markets. For Northern Ireland firms these markets included; the province of Northern Ireland; Great Britain; the Republic of Ireland; and outside the British Isles. Similarly, for the Massachusetts firms, the relevant market areas included; within Massachusetts; New England (excluding Massachusetts); the remainder of the USA; and outside the USA. While it is acknowledged that there are significant differences between the comparative regions in terms of population, economic activity, and geographical scale, this categorization is relevant because it represents the perceived potential trading areas/blocs as identified by the interviewees.

The home, or local market, is viewed by Northern Ireland firms as the province of Northern Ireland, with the home market for the Massachusetts firms being the state of Massachusetts, and in particular the Greater Boston area. Further, while the Northern Ireland firms perceive the Great Britain market as having political, economic and historical ties which assist in trading, a similar perception is held of the United States market by Massachusetts firms. Matched plants within both regions also identify a neighbouring market, namely the Republic of Ireland for Northern Ireland firms, and the other New England states for Massachusetts firms. Therefore while these markets may appear incompatible in real terms, if studied within the context of the behavioural environment (Kirk, 1963) then each of the comparative regions has strong links and has significant implications in attaining an understanding of the nature of production.

Table 7.1
Spatial distribution of outputs

	Northern Ireland	Massachusetts	Northern Ireland	Massachusetts
	Engineering (Percentage of sales)		*Food Processing (Percentage of sales)*	
Engineering				
NI/Mass	56.2	41.6	86.5	59.5
GB/USA	30.3	42.5	10.5	19.5
ROI/New England	6.8	12.2	3.0	12.5
Non-British Isles/Non USA	6.7	3.7	0.0	8.5

Source: Matched plant survey

Given the considerably smaller 'local' market area of Northern Ireland as compared to Massachusetts, it would have been expected that Northern Ireland firms would have developed their external markets to a greater extent than Massachusetts firms. This was not found, with the results (Table 7.1) illustrating that the distribution of output within the local market equated to 41.6 per cent of sales from the Massachusetts engineering firms, as compared to 56.2 per cent for the Northern Ireland firms.

Amongst the food processing matched plants the distribution of output was similar to that for the engineering firms. Again the Northern Ireland firms had a higher percentage of their output remaining within the immediate geographical area (86.5 per cent) as compared to the Massachusetts firms (59.5 per cent). The results support the view that the food processing sector relies to a much greater extent upon the local market than that of the engineering sector, due to the perishability of food products and low value to bulk ratios.

Table 7.2
Spatial distribution of raw materials and component parts

	Northern Ireland	**Massachu-setts**	**Northern Ireland**	**Massachu-setts**
	Engineering (Percentage of inputs)		*Food Processing (Percentage of inputs)*	
Engineering				
NI/Mass	43.5	47.1	51.2	53.8
GB/USA	43.5	45.6	42.7	40.0
ROI/New England	6.2	3.8	5.9	4.7
Non-British Isles/Non USA	6.8	3.4	0.0	1.5

Source: Matched plant survey

With respect to the source areas from which the matched plants received their inputs, again the dependence upon the local market area is found (Table 7.2). Unlike the distribution of outputs, engineering and food processing firms within both regions displayed a similar level of dependence upon the local market. For example, within the engineering matched plants, 43.5 per cent of inputs into Northern Ireland firms and 47.1 per cent of inputs into Massachusetts firms came from the local area. Similarly, for the food processing matched plants 51.4 per cent of the inputs into Northern Ireland firms and 53.8 per cent of inputs into Massachusetts firms came from the local area.

Whilst there was similarity in the percentage of inputs from the local markets into the Northern Ireland and Massachusetts engineering firms, this concealed differences within the engineering sectors. For matched plants producing metal products a similar proportion of both supplier parts and final products were located within the local area, with a similar proportion of goods being bought from, and distributed to the external markets. The Massachusetts electrical and electronics firms, unlike the other engineering sectors, obtained a much higher percentage of their inputs from within the local market (43.1 per cent) than their Northern Ireland matched plants (28.8 per cent). However, both the Northern Ireland and Massachusetts firms obtained a similar proportion of their inputs from the Great Britain (56.2 per cent) and United States (51.8 per cent) markets respectively.

The greater dependency upon the local market by Massachusetts electrical and electronic matched plants is reversed with respect to the food processing firms. Here the Northern Ireland firms depend upon the local market for 92.5 per cent of their inputs as compared to 45 per cent by the Massachusetts firms. A further 6 per cent of the raw materials for the Northern Ireland firms came from the Republic of Ireland and 1.5 per cent came from Great Britain. Conversely 34.3 per cent of inputs within the Massachusetts firms originate within New England with 20 per cent coming from the rest of the USA.

[1]In summary it can be concluded that the metal products and mechanical engineering firms within both Northern Ireland and Massachusetts were able to obtain approximately half of their inputs from the local market. In contrast, while 43.1 per cent of inputs into the electrical and electronics firms within Massachusetts came from the local area, the corresponding value for Northern Ireland firms was 28.8 per cent.

In terms of the impact of buyer-supplier relationships upon the ability of firms to introduce flexible production, it can be concluded that there is a similar spatial dependence upon suppliers and customers amongst the metal products and mechanical engineering matched plants. For electrical and electronics matched plants, the greater localization of suppliers within Massachusetts suggests that firms within this sector would be more capable of introducing JIT practices than the Northern Ireland firms. Yet overall, measures of the spatial distribution of outputs illustrate that a larger proportion of engineering and food processing products remain within the local area for Northern Ireland firms which would suggest that the Northern Ireland matched plants are better positioned to take advantage of JIT and other flexible production methods than the Massachusetts firms.

Manufacturing for stock

The elimination of buffer stocks is central to the JIT principles of production. Under Fordism while the core activities of production were revolutionized, inventory control was largely ignored. It has been stated that up to 30 per cent of production costs are attributed to poor inventory control and the warehousing which this demands (*Business Week*, 14 April, 1985:45). Buffer stocks have been associated with productive inefficiencies in terms of labour producing more than is needed at any point in time, 'idle time, surplus workers, excessive equipment capacity and insufficient preventive maintenance' (Sugimori et al., 1977). The primary aim of flexible production is to integrate resources more efficiently so that through the use of flexible technology (in conjunction with multi-skilled workers) firms may respond to demand in a manner which eliminates the need for buffer stocks.

Flexible production enables small firms to gain competitive advantages through economies of scope. The cost of technology may however be greater than the resources of small firms (Storey, 1994). Where small firms are unable to introduce flexible technology it is likely that production runs will be longer due to longer setting-up periods. Hence, where smaller firms are supplying to larger firms, who demand a rapid response rate to changing demands, this may culminate in the increase of buffer stocks as opposed to the reduction of them. Gertler also identifies this problem and states that 'although adopting a JIT/Kanban production system may reduce the need for buffer stocks in the factories of producer/assemblers, this may achieve little more than shift the location in which these suppliers (rather than assemblers) now stockpile instead' (Gertler, 1988: 427).

Where small firms are supplying to larger firms which are operating JIT production methods, or alternatively, where small firms are supplying to a market which is volatile in demand, unless the small firm can implement JIT principles, the extra costs of buffer stocks will render it uncompetitive relative to its competitors. Where firms and markets are demanding shorter lead-times, JIT enables small firms to respond both quickly and efficiently through the reduction of costs associated with buffer stocks.

Within the Northern Ireland and Massachusetts matched plants, the interviewees were asked whether they 'manufactured for stock'[2]. The findings demonstrated that the Northern Ireland firms had a greater propensity to manufacture goods for stock than the Massachusetts firms. Overall, approximately two-thirds of Northern Ireland firms stated that they manufactured for stock as compared to only half of the Massachusetts firms. However, differences existed with respect to the propensity of firms within each of the industrial sectors to manufacture goods/components for stock. With the exception of the metal goods sector, a smaller proportion of the Massachusetts firms manufactured for stock than the Northern Ireland firms (Table 7.3).

Regional disparities in the extent to which engineering firms manufactured for stock were less apparent within the food processing firms. Only 20 per cent of the food processing firms within both Northern Ireland and Massachusetts manufactured for stock, with the remaining 80 per cent, or 12 matched plants not undertaking production for 'stock'.

Table 7.3
The percentage of firms manufacturing for stock

	Northern Ireland	Massachusetts
	Percentage of firms	
Engineering		
Metal Goods	59.0	68.0
Mechanical Engineering	66.6	44.4
Electrical and Electronic Engineering	66.6	33.3
Food Processing	20.0	20.0

Source: Matched plant analysis

It is acknowledged that the implementation of JIT practices will be most effective where the proximity of suppliers to buyers/markets is closest. For example, Oberhauser (1990:218) states that, 'certain applications of flexible accumulation require spatial reintegration of production into regional complexes with easy access to suppliers, technical services and a combination of highly skilled and semi-skilled workers.' To some extent the small electrical and electronic engineering firms within Massachusetts conform to this association. Both within Northern Ireland and Massachusetts, the electrical and electronic engineering firms manufactured a smaller amount of goods for stock than any of the other engineering sectors. Yet, these firms also supplied a greater proportion of their goods to the local market. This would therefore support the existence of an association between a low inventory level and close proximity to markets.

The extent to which firms manufactured for stock and the distribution of goods for the engineering sector as a whole, fails to support the inventory-market association. For example, 56.2 per cent of goods manufactured by the engineering sector within Northern Ireland were distributed within the local market area, as compared to 41.6 per cent for the Massachusetts engineering firms. It would have been anticipated therefore that the Northern Ireland engineering firms would have had a lower proportion of firms manufacturing for stock than the Massachusetts firms. Yet this was not found, with 64 per cent of the Northern Ireland engineering firms manufacturing for stock as compared to 48 per cent within Massachusetts. This would therefore suggest that the Northern Ireland firms were less flexible (as characterized by the more rigid JIC principles of production) than the Massachusetts engineering firms[3].

The nature of subcontracting relationships

Subcontracting may be defined as 'performing some portion of work that has already been contracted for, implying at least a three-tier hierarchy…but it is also commonly used with respect to contracts drawn up between relatively equal firms - what we call simply buyer-supplier contractual relations' (Sayer and Walker, 1992:130). Large firms within the subcontracting hierarchy act as the focal point from which the vertical disintegration of production emanates. The control which the large (or core firms) have over the firms in the lower part of the hierarchy does not directly affect the legal ownership of these smaller independent firms. Instead, control is exerted through mechanisms such as 'cross-ownership, spatial proximity, information exchanges, and personnel sharing' (Florida and Kenney, 1990:243, Altshuler et al., 1984). This form of control has advantages for the core firms in the form of; 'accelerated turn-over time, reduction in the organic composition of capital, improved resistance of the integrating firm to cyclical market conditions, increased flexibility for the mother plant and the inequality of economic power…between integrating and integrated firm…reduction of the cost of labour, improved productive use of labour time...' (Swyngedouw, 1991:329/330).

This would suggest that the small independent firm was being peripheralized relative to the core firm. The small firm in absorbing the risks and costs of production also protects the larger firms against demand variability and uncertainty (Moulaert and Swyngedouw, 1989). Where large firms within a region are operating flexibly, the restructuring and rationalization of their production will lead to the widespread use of subcontracting. This will therefore permit large firms to focus more closely upon other activities, such as, research and development, design and quality management etc. The extent to which small firms undertake subcontracting relationships would therefore reflect the pervasiveness of flexible production strategies.

Japanese small firms formed through company spin-offs, act as a protective periphery to the larger core firms, in providing numerical flexibility. Conversely, disintegration of production and the formation of subcontracting relations within the West are mainly the result of market forces, with firms maintaining their independence (Florida and Keeney, 1990). Yet, where small firms are highly dependent upon one or two contracting firms, then despite independence of ownership, considerable control and influence will be exerted over the subcontractor. Additionally, over-dependence upon one or two contractors will increase the vulnerability of small firms to changes in demand.

Results from the matched plants (Table 7.4) showed that 70 per cent of the small engineering firms within Northern Ireland acted as subcontractors to other firms, while the corresponding figure for Massachusetts was 93 per cent. This would suggest that for the engineering firms inter-firm linkages were more important amongst the Massachusetts firms.

Table 7.4
Percentage of firms involved in subcontracting relationships by industrial sector

	Northern Ireland	Massachusetts
	Percentage of firms	
Engineering		
Metal Goods	75.0	91.6
Mechanical Engineering	44.4	88.8
Electrical & Electronic Engineering	88.8	100.0
Food Processing	73.3	80.0

Source: Matched plant analysis

The use of subcontracting relations varied across the engineering sectors. Yet irrespective of the industrial sector, Massachusetts small engineering firms acted as subcontractors to a much greater extent than the Northern Ireland engineering firms. Within both regions a greater proportion of electrical and electronic firms acted as subcontractors than within either the metal products or mechanical engineering sectors. This reflects the nature of the products being produced within these sectors. For example, a number of the electrical and electronic firms manufactured printed circuit boards which only obtained their use value through their incorporation into other products such as computer systems, televisions, radios, engines, alarm systems etc.

The out-sourcing of work is not solely the domain of the larger firms, but indeed may be used quite effectively by small enterprises. For example, within one of the Northern Ireland firms which manufactured specially designed tubes for use in the installation of refrigeration equipment, financial losses at the end of 1984 led to the installation division of the firm being closed. In making these workers redundant, the company, acknowledging that installation services were essential to the firm, established a contracting relationship with the employees. The firm gave each of the workers a van and the tools which were essential to the installation of refrigeration and heating systems, and with this an 'unwritten' code of practice was established that these workers would work for the firm whenever needed, yet were simultaneously independent of the firm. Thus, the firm externalized the costs and responsibilities which the employment of these people generated, while simultaneously maintaining an effective service from them whenever needed.

The results suggest that, within the environment of the matched plants there was greater flexibility within Massachusetts, as manifest by a higher proportion of engineering firms involved in subcontracting. It is possible, however, that where the small firms are subcontracting a large proportion of their total production for only one or two firms, then this may limit the independence and autonomy of the small firms. Therefore by looking at the number of firms which these small firms subcontracted for, in conjunction with the incidence of subcontracting, an impression of market vulnerability is attained. Within Northern Ireland, those firms which took part in subcontracting did so with an average of 3.8 firms, the corresponding number of firms within Massachusetts being 8.3. This would suggest that the Massachusetts firms which were involved in subcontracting, were therefore less vulnerable to changing market demands within the contracting firms, as compared to their Northern Ireland counterparts.[4]

Amongst the food processing firms, similarity was found in the incidence of subcontracting relationships within the sample of Northern Ireland and Massachusetts firms. Within Northern Ireland 73.3 per cent of the matched plants participated in subcontracting, as compared to 80 per cent within Massachusetts. A significant difference does however exist with respect to the number of firms for which the matched plants act as subcontractors. Within Northern Ireland subcontracting firms on average manufactured for 2.3 firms, while the corresponding number of firms within Massachusetts was 4.6. While the maximum number of firms for which a Northern Ireland food processing matched plant manufactured was five firms, within Massachusetts one of the food processing firms manufactured for 12 other firms. Therefore on average, the Massachusetts food processing matched plants were subcontracting for a larger number of firms than the Northern Ireland matched plants.

Not only did a higher proportion of food processing firms take part in subcontracting, than the engineering firms, but on average, within both regions, subcontracting relationships within the food processing firms were undertaken with a smaller number of other firms. The food processing differs from the engineering sector in terms of the stability of the market. Overall, within the developed world the demand for food is relatively static[5], with innovations only being able to displace other sectors of the market. Conversely, within the engineering industry, new innovations can create substantial new demand (although in some cases this entails the displacement of some other sectors). Large firms within the food processing sector can therefore benefit from economies of scale and encourage customer loyalty through advertising. Conversely, small firms which are financially constrained and frequently unable to produce an output of substantial size to justify the use of advertising on a broad geographical scale, are marginalized in the selling of their output. Thus, the large firms increase their incidence of out-sourcing products to smaller firms in an attempt to externalize the risks of production which may result from changing market conditions, and as such the smaller firms become dependent on the large firms, as an outlet for their products. Yet, the dominance of the food processing sector by a small number of large firms means that small firms will be highly dependent upon a small number of larger firms, as compared to the engineering sector (Beaumont, 1982)[6].

As the matched plants were all independently owned, the contractual firms did not have a legal hold over the subcontractors. Yet, for both sectors contractual arrangements had implications for the small firms in terms of conditions, or alliances of production. Some of the food processing firms within both regions had marketing agreements in that they manufactured to pre-determined criteria with the product being sold under the 'branded' name of the contractual firm. Amongst some of the Massachusetts engineering firms which took part, subcontracting 'technology-use' agreements were being used, whereby technology used by the contracting firm was licensed to the small firm.

These subcontracting agreements have been defined as 'relational' in that they represent an 'ongoing relation of exchange, interaction and mutual development between two or more firms... and some degree of involvement by one firm in the production process of another, such as design specifications, providing materials and machinery, technical assistance, financial assistance and on-site quality control' (Sayer and Walker, 1992:130).[7]

The extent to which exchange, interaction and mutual development were evident within the subcontracting relationships varied considerably both by region and sector. For example, on four occasions within the food processing firms within Massachusetts, personnel from the contractual firms were present within the factory, with on one occasion, the person arriving unexpectedly during the course of the interview, to inspect the standards of both the quality of the products and the hygiene criteria being implemented. While it is acknowledged that such co-operation did exist within the sample of Northern Ireland firms, these 'inspections' were more formalized, regular and planned, with firms receiving prior warning of visits from contracting firms.

The extent of relational subcontracting amongst the engineering firms within Massachusetts was more developed than for the Northern Ireland subcontracting firms. The engineering sector differed from food processing in that there was less involvement by the contracting firm in the production process. The contracting firm in specifying the design and manufacturing criteria with the small firms merely inspected the final product and ensured that all contractual arrangements, such as delivery times, were met. While this form of interchange was most commonly found amongst the Northern Ireland subcontracting firms, the Massachusetts engineering firms were involved in greater exchange, interaction and mutual development. For example, it was not uncommon to find technology-use agreements and personnel working together to improve product design and capability.

In summary, although the incidence of subcontracting relationships was only slightly higher amongst the Massachusetts engineering and food processing firms, the nature of contracting relationships varied between the Northern Ireland and Massachusetts firms. For example, within the Northern Ireland engineering firms, most of the subcontracting orders were inconsistent in both demand and volume. Therefore, the contracting firms were using the small firms as a protective periphery in providing them with components as their own demand changed. These small engineering firms were therefore constrained by the unstructured and unreliable nature of subcontracting orders. It could be inferred from this that the greater uncertainty of demand within the sample of Northern Ireland engineering firms explains the greater extent to which these firms manufacture for stock (see Table 7.1).

In contrast, the nature of subcontracting within the Massachusetts engineering firms was more structured, with relationships being nurtured between firms as evident through the exchange of personnel and technology. In particular, all of the production within the electrical and electronic engineering firms within Massachusetts was for other firms, mostly located within the local market area, and especially along Route 128. For these firms, the greater reliability upon a more developed form of 'relational' subcontracting meant that firms could incorporate such orders within their production planning and therefore reduce the necessity to manufacture for stock.

As firms within both regions distributed a similar proportion of their output into the local markets, the location of buyers cannot explain the differences in subcontracting. The differences may be accounted for by other factors which have previously been alluded to. These include the manufacturing efficiencies of the firms (in terms of productivity, performance and quality), and the inherent nature of subcontracting relationships, whereby firms are seen less as a 'protective periphery' in absorbing the risks and costs of demand variability, and instead are viewed as business units with expertise and strengths which other firms can benefit from.

The incidence of R&D activity within the matched plants

Florida and Kenney (1990) highlight that within Japanese firms R&D is an integral part of manufacturing with linkages developing between the innovative phase of production and that of manufacturing. Through the overlapping of functions between the innovative and manufacturing phase, a variety of workers are involved in various stages of product development, with these workers attaining skills which would not have otherwise been possible. Traditionally, under Fordist principles of production R&D was perceived to emanate from large research and development departments, from which the idea would finally be manufactured on the shop floor. It was therefore believed that R&D activities would be located in core areas which experienced higher rates of regional growth, while the production activities would be located in peripheral branch plants (Thwaites, 1983).

The linear model of R&D has however been challenged widely (Piore, 1968; Freeman, 1982; Rees et al., 1985) with the undertaking of R&D being seen more as an integral process which crosses occupational boundaries. It has been found that 'good ideas must be transformed into prototypes, then incorporated into operational production systems; this requires the labor and intelligence of skilled craftsmen, production managers, salespeople and the like' (Storper and Walker, 1989:103).

For the small firm, flexible technology enables them to attain competitive advantages based upon economies of scope (Piore and Sabel, 1984). Product differentiation will therefore be central to the maintenance of competitive advantages (Porter, 1980). Where the division of labour is less fragmented amongst small firms it will be paramount that the workers are functionally flexible (Atkinson, 1985). Through multi-skilling and job enhancing, workers within the small firms will be transferable throughout the workplace and through working interdependently establish an innovative culture which 'makes much less tenable the assumption ... that invention is the key element which automatically initiates a chain reaction' (MacDonald, 1983:31, as quoted in Storper and Walker, 1989:103).

As functional flexibility is central to the undertaking of R&D in the flexible firm, then the results from chapter five which demonstrated a higher overall level of functional flexibility within the Massachusetts firms, is important for the innovative capacity of the matched plants. The comparative advantage of higher educational qualifications amongst the Massachusetts employees was accentuated by the higher levels of per capita investment in training. These firms had eliminated job demarcations to a greater extent than within the Northern Ireland firms, and there was a greater propensity for team working as evident through the interdependent relationships between workers. It would be anticipated that the Massachusetts firms would therefore have been more capable of undertaking R&D functions.

Regional and sectoral differences in the percentage of matched plants which were actively involved in R&D are illustrated in Table 7.5. For the engineering sector, 30 per cent of the Northern Ireland matched plants were undertaking R&D as compared to 46 per cent within the Massachusetts sample. Examining the incidence of R&D by industrial sector presents greater variations between the samples.

If the extent of R&D is analyzed by industrial sector an interesting and significant picture emerges in terms of the competitive nature of the engineering sector within both regions. Firstly, within the older and less technologically advanced metal products sector, the comparative extent of R&D is similar between the two regions. However, within both the mechanical engineering and electrical and electronics sectors, the Massachusetts matched plants were utilizing R&D more intensely than their Northern Ireland counterparts, that is, by 22 per cent and 33 per cent respectively. This therefore conforms to the findings on functional flexibility and suggests that the presence of multi-skilled staff had an important impact upon the capability of the small engineering firms to undertake R&D.

Table 7.5
Percentage of firms actively involved in research and development

	Northern Ireland	Massachusetts
	Percentage of firms	
Engineering		
Metal Goods	33.3	33.3
Mechanical Engineering	44.4	66.6
Electrical and Electronic Engineering	11.0	44.4
Food Processing	0.0	13.3

Source: Matched plant analysis

The proportion of matched plants undertaking R&D was significantly different from the engineering sectors. None of the food processing matched plants in Northern Ireland undertook R&D in the 12 months prior to the interviews. Amongst the Massachusetts firms, only two of the 15 matched plants had performed R&D functions. Where R&D did occur within the matched plants it was performed within a formal research and development department. The linear model of innovation was therefore more applicable to the food processing firms than the engineering firms. As outlined in chapter five, the food processing firms also had more rigid job demarcations than the engineering sector, with a higher proportion of workers operating independently of each other. Further, within the food processing sector there was greater evidence of a peripheral work force as seen through the significantly higher proportion of employees without formal qualifications. These employees acted as production line workers, in performing rote tasks which required low levels of skill.

The differences between the engineering and food processing firms in the extent to which R&D was undertaken, suggest that it is a combination of the capability of workers as well as the organization of production which determine the nature of R&D. Where production is low-skilled, and characterized by standardized tasks with little interdependence between employees where R&D is found it will be within a separately identifiable department. Conversely, where production is semi-skilled or highly skilled, with employees being encouraged to work together in teams it is more likely that R&D will be an integral part of the firms organizational strategy.

While flexible technology may enable small firms to set-up machines quickly, and therefore benefit from economies of scope, it is unlikely that they will be unable to develop anything more than an *ad hoc* approach to R&D. The majority of small firms will not have the capital and labour resources to establish a formal R&D department and therefore if R&D is to occur, it must be informal[8]. Yet, this can only be attained through the nurturing of employees' skills and the creation of an environment whereby exchange and development exists between workers.

Within the small food processing firms approximately 53.3 per cent of the firms interviewed within both regions indicated that 'size' was a restrictive element in their attainment of growth[9] . Additionally, 33 per cent of the Massachusetts firms and 13.3 per cent of the Northern Ireland firms stated that they found their 'market dependency' upon contracting firms as a significant limiting factor. This market dependency was restrictive in that the small firms were highly dependent upon the contracting firms and where new products were introduced these were developed by the contractors and merely implemented within the small firm. Hence, the small food processing firms, while legally and financially independent from the larger contractors (upon which the majority of their market share depended) were tightly bounded by the products and processes which these firms specified in their contracts of production.

Similarly, none of the engineering firms undertaking R&D in either Northern Ireland or Massachusetts had a formal R&D department. Amongst a number of the Massachusetts firms which were undertaking subcontracting within other firms and where this was characterized by significant interaction and exchanges of information and technology, R&D was pursued more intensely. Therefore the subcontracting relationships amongst the Massachusetts firms were providing mutual benefit with R&D stimulating product and market diversification for both the contracting and subcontracting firms alike[10].

The incorporation of quality control within the production process

Fordist principles of production were based upon Taylor's perception that there was 'one best way' for production, with workers being organized to minimize and eliminate the need for them to make judgements. Quality was therefore assigned to a separate department with quality not being monitored on the assembly line. Although defective parts may have been observed by production workers, the nature of their jobs and the way pay was linked to output, did not permit or encourage them to remove physically defective parts, or inform other workers of possible errors in production. Defective parts and goods therefore by-passed assembly line workers. Yet the quality control department did not inspect every product, but instead monitored quality through the random selection of goods. A number of defective products were therefore undetected in the quality control process and concealed in the buffer stocks. Within this Fordist system of quality control there is therefore an under-utilization of resources; firstly, the work remit of shop-floor employees excludes the detection and elimination of errors, and secondly, through the utilization of a separate department upon the shop-floor dealing with the monitoring of quality control, a large proportion of productive work is relinquished.

Although a number of management and social scientists challenged this form of worker control (McGregor, 1960; Lewin, 1951; Mayo, 1949) and instead argued for greater self direction and self control amongst production workers, examples of greater employee involvement during the 1960s and 1970s were rare (Fenton-O'Creevy, 1994). Yet, the Flexible Specializationists (Piore and Sabel, 1984; Sayer, 1985; Murray, 1988, 1989) argued that greater involvement would grow naturally out of the technologically-led organizational innovations in the flexible firm, and this would provide employees with some control and determination. Beyond the Flexible Specialization school, growing international competition and declining productivity were encouraging companies to re-examine traditional production and management methods. The crisis of industrial relations within the automobile industry and the relative success of Japanese firms made industrialists look for ways to improve their productivity. Greater employee involvement in the form of quality circles was identified as a way to improve industrial relations; encourage employee commitment; raise productivity; and reduce the wastage of resources by attaining zero defects (Jurgens, 1989).

Through increased employee involvement, each production worker now takes an active role in monitoring product quality and the need for a separate department dealing with quality control is eliminated. It is argued (Jurgens, 1989) that although greater employee involvement and the formation of quality control circles was oriented towards quality goods, groups of workers have become increasingly involved in workflow organization and questions of work safety. Quality circles are therefore more concerned with process, as opposed to product innovations, and most importantly, the elimination of wasteful activities (Kumazawa and Yamada, 1989).

Quality control circles are defined as systems through which the quality of production is monitored by all employees working within small integrated teams upon the shop-floor, with each team being accountable to the other teams within the production process. Through this system workers no longer operate independently of their colleagues, but rather interact with them on a continuous basis as manifest through the establishment of production schedules, setting of machines, operation of machines, maintenance of machines and the inspection of production. Therefore each worker as part of a team becomes accountable not only to the management, but also to their colleagues.

Kumazawa and Yamada (1989:117) illustrate that 'when a quality control circle tackles a technical problem, it is usual that the circle conducts a systematic survey of its members' performance, using various industrial engineering methods including time and motion studies'. An examination of the effects of quality control circles within Sumitono Metal Industries has demonstrated that they led to substantial improvements in production. Levels of efficiency were raised by 26 per cent; the quality of products was improved by 22 per cent; the costs of production decreased by 15 per cent; saving with respect to the use of energy equated to 12 per cent; managerial efficiency was raised by 7 per cent; safety was improved by 5 per cent; and all 'other' aspects improved by 13 per cent. (Swyngedouw, 1991:318). From this it was concluded that quality control circles aid in 'increasing productivity, reducing 'dead' labour time and ameliorating efficiency, (and thus) labour contributes actively to improving surplus value production gains and thus, to intensifying its own exploitation'. Tomaney (1990:34) also accentuates this concept of labour increasing its own intensification through the acceptance of quality control circles in that he sees participation within quality control circles as being coerced rather than voluntary, with such processes 'ensnaring' workers within a system of intensification. Hence, Tomaney concludes that quality control circles are directly related to process as opposed to production innovations, with the presence of such being manifest through the elimination of 'wasteful activities'.

Where firms are implementing flexible principles of production it would be anticipated that in the monitoring of quality, this would be an integral part of the production process as opposed to being performed by a separate group of employees at the end of the production process. Further, as the incorporation of quality control within the production process, in the framework of quality control circles, is related to the elimination of wasteful resources, it would be expected that this would lead to a lower reject rate than found under traditional Fordist principles.

Table 7.6
The percentage of firms incorporating quality at the end, or during the production process

	Northern Ireland		Massachusetts	
	Percentage of firms			
	At End of Production	*During Production*	*At End of Production*	*During Production*
Engineering				
Metal Goods	66.6	33.3	0.0	100.0
Mechanical Engineering	78.0	22.0	11.1	88.8
Electrical & Electronic Eng.	33.3	66.6	0.0	100.0
Food Processing	60.0	40.0	0.0	100.0

Source: Matched plant analysis

Within the matched plants, significant differences were found in the way quality control was incorporated within the firm (Table 7.6). Firstly, for the metal goods and mechanical engineering firms within Northern Ireland, quality control was mainly performed at the end of the production process, where quite often it is too late to alter the products, and where re-working is impossible. It would be anticipated that there would be a high level of wastage amongst these firms. Conversely, for the Massachusetts firms within each of the engineering sectors, the method of quality control which predominated was that of continuous monitoring and checking of products throughout the production process, both by the shop-floor workers, and by managerial or supervisory staff. Where errors are detected at an early stage in production, products can be reworked and a lower reject rate achieved. These findings show that the Massachusetts engineering firms operated a system of quality control which was an integral part of the production process. Conversely the Northern Ireland firms conformed more closely to traditional principles of production organization, with quality control being performed at the end of the production process, and by a separate group of employees.

Within the sample of Northern Ireland food processing firms only 40 per cent employed a policy of product checks during the production process equating to six firms. Consequently, nine firms, or 60 per cent of the Northern Ireland sample, left quality control to the end of the production process. In contrast, all of the 15 small food processing firms within Massachusetts reported that they monitored production throughout all the relevant stages.

The findings for the food processing industry are significant in that they demonstrate regional as opposed to sectoral differences within the sample of firms. While the Massachusetts firms appear to be reducing the rigidities of Fordist forms of production, and instead are adopting working practices which facilitate both quantitative and qualitative advances in production the Northern Ireland firms have a more clearly defined internal division of labour. The Massachusetts firms demonstrate that JIT principles of production are not only applicable to the more technologically advanced industries, such as the electrical and electronics sector, but they may also be applied to traditional industrial sectors such as metal goods and food processing.

The results demonstrate that significant differences existed between the Northern Ireland and Massachusetts firms in the way quality control was incorporated within the production process. It would be expected that where quality control was performed as part of the production process, then this would result in lower reject rates, as compared to where quality control was performed at the end of production. A close correlation was found between the reject rate of production and the way quality control was incorporated. For the sample of engineering firms, the average reject rate was 5.13 per cent in Northern Ireland and 1.78 per cent in Massachusetts. While the difference in reject rates is high for the mechanical engineering and electrical and electronic firms, it is slightly smaller for the metal goods firms, where the Massachusetts average reject rate was approximately 50 per cent that of the Northern Ireland rate. Conversely, for the mechanical engineering and electrical and electronics firms, the Massachusetts average reject rate corresponded to 28 per cent and 25 per cent of the Northern Ireland rate respectively. A positive association therefore exists between the manner in which quality control is incorporated into the production process and the reject rate of production within the engineering firms.

Within the food processing firms, again there was a significant difference between the overall reject rate within Massachusetts as compared to Northern Ireland. For example 33.3 per cent of the Massachusetts firms stated that they had a 'zero' reject rate with only 20 per cent having a reject rate between 0.1 and 1 per cent. Therefore, while 53.3 per cent of the Massachusetts small food processing firms had a reject rate below 2 per cent, all of the Northern Ireland firms had a reject rate in excess of 2 per cent.

The mean reject rate amongst Northern Ireland food processing firms was 2.8 per cent, while the corresponding figure for the Massachusetts firms was 1.7 per cent. Although the differential between the mean of the two regions is smaller than that found for the engineering industries the lower reject rate within the Massachusetts firms is still significant. Indeed, as was found for the engineering firms, and is here further accentuated, there is a direct and positive correlation between the nature of the quality control and the reject rate of production.

The age and technological sophistication of capital

In examining the extent of flexibility in production the technological determinists (Piore and Sabel, 1984; Sayer, 1985; Murray, 1988, 1989) argue that the introduction of flexibility will occur as a result of the introduction of flexible technology. Yet the age of capital is also significant in that, theoretically, newer equipment will be more efficient and therefore positively affect productivity. On balance, the lower the age of the capital equipment, the better the performance of firms (McFerran et al., 1996).

Technological innovations can be subdivided into two categories, namely those which 'have enabled firms to organize their production in more flexible ways and which may be described as the 'techno-structure' of flexibility ... and those innovations, again often microprocessor based, whose main effect has been to increase productivity and output rather than increase flexibility per se.' (Morris, 1988:309). The introduction of new technology lies at the centre of the new forms of flexible production. For example, the use of numerical controls is a central element in the adoption of flexible machines. Numerical controls comprise a system in which specified actions are determined by the insertion of numerical data into the electronic systems of machines. The advantage of this system lies within the flexibility which it provides, whereby through the manipulation of data, production can quickly and efficiently be switched from one component to another. This system is also believed to be more accurate than traditional systems along with displaying a higher cost efficiency due to the reduction and virtual elimination of extensive set-up times (see Allen and Massey, 1988).

While numerically controlled machines are important in the production process, it is the attainment of a Flexible Manufacturing System (FMS), which is viewed as the goal of the flexible firm. The implementation of a FMS consists of, 'combining a number of flexible machines into an integrated aggregation (usually linked together and coordinated by computer) ... (whereby) one can form either small scale manufacturing cells or larger clusters of computer integrated manufacturing (CIM)' (Gertler, 1988:420).

The ethos of flexible machinery is perceived as an integral part of the flexible firm. Technology can be used to integrate design (CAD) with manufacturing

(CAM) and distribution, to attain economies of scope as well as those of scale. Therefore, 'the key distinction between flexible technologies and traditional mass-production techniques is that what the machine actually does is programmed in via computer software rather than built into the machinery at the outset. This means that a given machine can be used in the manufacture of a range of product types and configurations; hence its flexibility in contrast to fixed-purpose dedicated machinery. Significantly, the existence of machines that can perform several functions, or variations of a given function according to product configuration, allows shorter runs and a mix of product types on the line without sacrificing economies of scale. Economies of scope can be exploited across a range of product types while aggregate production volumes can remain quite high in principle' (Schoenberger, 1988:252).

It would be anticipated that those firms with the youngest and most advanced capital equipment would also be those firms which performed better (McFerran et al., 1996) and had introduced flexible principles of production to a greater extent. In comparing the capital equipment between the matched plants, the presence of 'advanced' machinery, the perceived percentage utilization of machinery, and the average number of breakdowns within the production process were examined.

The percentage of machinery for the engineering and food processing firms which was under five years; between five and ten years, and over ten years is illustrated in Table 7.7. While it was anticipated that the small engineering firms within Massachusetts would have had a higher percentage of 'new' machinery, that is, capital under five years of age, in line with their greater use of Just-In-Time principles, this was not found. Instead, the Northern Ireland firms appear to have a slightly higher percentage of 'new' machinery than their Massachusetts counterparts. Yet, the percentage differences between the engineering firms are slight.

Amongst the food processing firms the Massachusetts matched plants had 35 per cent of their capital under five years of age as compared to 23 per cent for the Northern Ireland matched plants. Little difference is however found for the food processing firms in the percentage of capital under ten years, and over ten years of age, between the two regions. These results show that there was not a significant difference in the age of capital between the sample of firms in Northern Ireland and Massachusetts. Neither of the firms would therefore have a 'performance advantage' as measured by the age of the capital stock.

Table 7.7
The average age of capital

	Northern Ireland	Massachusetts
	Percentage of firms	
Engineering		
<5 years	28	22
5-10 years	30	33
10+ years	42	45
Mean Age	8.2	8.6
Food Processing		
<5 years	23	35
5-10 years	39	25
10+ years	38	40
Mean Age	8.25	7.7

Source: Matched plant analysis

Examining the capital stock in terms of the percentage of firms which had introduced computer controlled or other technologically advanced machinery (e.g. robotics) demonstrates disparities between the matched plants (Table 7.8).

For the sample of engineering firms, 66 per cent of the Northern Ireland matched plants had introduced advanced technology, whereas the comparative figure for Massachusetts was 83.3 per cent. Examining these results by industrial sector illustrates that Massachusetts firms, within each engineering sector, had a higher presence of technologically advanced machinery within their firms. Therefore, although the Massachusetts firms had a lower percentage of their machinery 'under five years' of age, they had a higher proportion of technologically advanced machinery.[11]

Table 7.8
The percentage of firms with technologically advanced capital

	Northern Ireland	Massachusetts
	Percentage of firms	
Engineering		
Metal Goods	59.0	78.0
Mechanical Engineering	66.6	92.0
Electrical & Electronic Engineering	78.0	92.0
Food Processing	0.0	13.3

Source: Matched plant analysis

The incidence of technologically advanced machinery was considerably less amongst the food processing firms in both Northern Ireland and Massachusetts. Within Northern Ireland none of the sampled firms had introduced such technology within the production process, with only two of the fifteen firms within Massachusetts having technologically advanced machinery within their capital stock.

While the age of capital and the incidence of technologically advanced capital are indicative of the productive capacity of the firms, and their capacity to be flexible, they do not illustrate the extent to which production capacity is being used. The capacity utilization of machinery within the matched plants was obtained (Table 7.9). Low capacity utilization rates are likely to equate to high capital costs per unit of output and may have an adverse effect on productivity, profitability and return on assets.

Perceived levels of capacity utilization demonstrate considerable comparability between the matched plants. For example, capacity utilization amongst Northern Ireland metal product firms is 69.8 per cent as compared to 67.9 per cent within the Massachusetts matched plants. Amongst the mechanical engineering firms a significant disparity exists with capacity utilization within the Northern Ireland firms being 67.4 per cent, with the corresponding Massachusetts utilization being 78.6 per cent. The largest disparities are found within the electrical and electronics sector with the

Northern Ireland firms utilizing their machinery by 70.7 per cent while the Massachusetts capacity utilization is only 57.9 per cent.

Table 7.9
Capacity utilization rates within the matched plants

	Northern Ireland	Massachusetts
	average capacity utilization (Percentage)	
Engineering		
Metal Goods	69.8	67.9
Mechanical Engineering	67.4	78.6
Electrical & Electronic Engineering	70.7	57.9
Food Processing	72.0	72.5

Source: Matched plant analysis

The levels of capacity utilization within the food processing matched plants were similar for both Northern Ireland (72 per cent) and Massachusetts (72.5 per cent). The majority of firms within the food processing industry are operating in the short-term, serving the immediate needs of the market. Due to the short shelf-life of processed products it is impossible for such firms to create buffer stocks, and therefore they must utilize some of the principles of Just-In-Time as an inherent part of production. Accordingly, comparability in capacity utilization between the regions is a consequence of the need to operate capital at sporadic intervals, as opposed to continuous production runs.

It would be anticipated that where a firm demonstrated organizational flexibility, as evident through; the elimination of lines of demarcation; interdependence of workers in self-managing, autonomous work teams or quality control circles; then skill enhancement and multi-skilling of workers would lead to the prevention of machine down-time through breakdowns (Sayer and Walker, 1989; Fenton-O'Creevy, 1994). Findings from the matched plants demonstrated that for each of the engineering sectors the Massachusetts firms had a lower number of capital breakdowns (Table 7.10). The Massachusetts food processing firms also had fewer mechanical breakdowns (2.3) than the Northern Ireland matched plants (3.7).

Table 7.10
Average number of machine breakdowns

	Northern Ireland	**Massachusetts**
	Average number of breakdowns	
Engineering		
Metal Goods	2.75	2.5
Mechanical Engineering	3.0	2.0
Electrical & Electronic Engineering	3.7	1.3
Food Processing	3.7	2.3

Notes: The average number of breakdowns refers to the month prior to the interview.

Source: Matched plant analysis

This would suggest that where the reduction in the lines of demarcation exist, in conjunction with the development of interactive relations between employees - most notably in the form of quality control circles - then capital equipment can be maintained and preventative measures taken to reduce the loss of productive time through capital breakdowns. Yet most importantly, where workers are trained and skilled to be functionally flexible throughout the firm, across occupational boundaries, then a policy of 'machine maintenance' can be established and hence potential faults can be rectified prior to production being halted.

In visiting the firms it was apparent that where team working and quality control circles were used - most frequently within the Massachusetts firms - then an integral part of this was the continual cleaning and maintaining of machinery by each employee. Conversely, within the Northern Ireland firms just as the use of a separate department for quality control existed, so a separate department for the maintenance of machinery also existed. Workers within these firms therefore, merely reported faults which arose and did not take any preventative measures to eliminate potential problems. One manager stated that his workers would not report a problem if they foresaw one, for in the event of a machine breaking, there would be a period of time whereby production would have to cease, and thus the workers would get a break. However, if they tried to prevent the breakage this would eliminate their 'free-time'. This view therefore accentuates the perception of Just-In-Time principles leading to the intensification of labour and demonstrates that it is only through the voluntary acceptance by workers of greater employee involvement that team-working can exist.

Results from the matched plants concerning the age and technological sophistication of machinery display considerable similarity between the regions. Although disparities were small, some differences were apparent. For example, overall, the Northern Ireland firms had a higher percentage of machinery under five years of age than the Massachusetts firms. However, the Massachusetts firms employed a higher percentage of technologically advanced equipment in the form of numerically controlled machinery and robotics. Of the numerically controlled equipment, CNC lathes and cutting equipment were most apparent along with Computer Aided Design (CAD) being frequently found where changing designs was an important element in the manufacturing process.

Of greater significance than the age and presence of technologically advanced machinery were the findings concerning the utilization of machinery and the manner by which the capital stock was maintained. The results support previous findings that the Massachusetts firms had adopted Just-In-Time principles of production to a much greater extent than the Northern Ireland firms. For both the engineering and food processing sectors employees within the Massachusetts firms played a key-role in the maintenance of machinery. Within the Massachusetts firms employees were expected to clean, oil, grease, sharpen etc. the capital equipment which they were using whenever they had 'free-time'. Conversely the Northern Ireland employees were in the majority of cases not expected to perform such a function, with a maintenance technician often being employed for this sole purpose. Hence, overall the preventative measures as employed by the Massachusetts firms were more profitable than those employed by the Northern Ireland firms, as measured by the number of breakdowns which had occurred.

Summary

This chapter has extended the analysis of flexibility beyond that of labour, to the production process within the matched plants. While indicators of flexibility in production are more elusive that that for labour, they are founded upon the aim of creating high quality products with the maximum level of functional efficiency. This therefore has implications for inventory control; inter-firm relationships; quality control; and the efficient use of capital.

The findings demonstrated that the Northern Ireland engineering firms had a greater propensity to manufacture for stock than the Massachusetts firms, with this disparity being most evident for the electrical and electronic matched plants. Yet, a similar proportion of the food processing firms within both regions manufactured for stock, suggesting a sectoral bias in the results with manufacturing for stock being less applicable within the food processing sector.

A slightly higher proportion of the Massachusetts engineering firms took part in subcontracting. This suggests that for the engineering firms, inter-firm linkages were more important within the Massachusetts sample. Of greater importance was the extent to which exchange, interaction and mutual development were found in the subcontracting relationships. Within the Northern Ireland sample, small firms acted as a protective periphery within subcontracting relationships. These firms were found to be constrained by the unstructured and unreliable nature of subcontracting orders. In contrast, the Massachusetts firms experienced more 'relational' subcontracting with mutual development being apparent in the undertaking of R&D.

Differences were most evident between the engineering firms in Northern Ireland and Massachusetts, in the manner quality control was incorporated within the production process. Amongst the Northern Ireland engineering firms, quality control was characterized by a strong internal division of labour with separate employees being responsible for the monitoring and inspection of quality, predominantly, at the end of the production cycle. Conversely, the majority of Massachusetts firms undertook quality control as part of the production process, with employees operating within quality control circles, providing continuous monitoring and the identification of defective parts. The organizational differences in quality control were manifest in the Massachusetts firms having significantly lower reject rates of production than the Northern Ireland firms. Therefore amongst both the Massachusetts engineering and food processing firms, the greater flexibility of labour in monitoring quality led to a reduction in the wastage of resources.

Piore and Sable (1984) argue that it is the introduction of technologically advanced machinery which will lead to increased flexibility within the firm. Analysis of the age and technological sophistication of capital did not however support this view. Comparability was found between the regions in the age of capital, with Northern Ireland firms having a slightly higher proportion of capital under five years of age. Further, a similar proportion of the firms had introduced technologically advanced machinery, although this was slightly higher within the Massachusetts sample. No significant differences were found in the productive utilization of capital between the regions, with the exception of the electrical and electronics firms, for which the Northern Ireland firms were using 70.7 per cent of their capacity as compared to 57.9 per cent within

Massachusetts. Massachusetts firms were utilizing their labour more efficiently in maintaining machinery than the Northern Ireland firms, and this resulted in a lower average number of capital breakdowns within the Massachusetts plants.

The results therefore demonstrate that there was greater utilization of Just-In-Time principles of production within the Massachusetts firms. Allowing for disparities between the two regions in terms of their capital stock, such differences do not adequately provide an explanation of the higher incidence of Just-In-Time within the Massachusetts firms.

It has been stated that; 'Advanced automation, such as flexible manufacturing systems has the greatest effects on productivity where it is applied to a production system that is already rationally organized; otherwise it is likely to perpetuate the inefficiencies of the old technology and working practices, as has been found with many major new technologies, including computer-integrated manufacturing and office automation' (Sayer and Walker, 1992:173).

This would suggest that while technological advancements may be useful in assisting the introduction of flexible working practices within the production process, on their own they merely serve to reinforce the rigidities and constraints of the former system. Therefore; 'a major shift in the underlying technology of production is also likely to be associated with changes in the technical and social division of labour, including the reorganization and recomposition of tasks and the elaboration of new relationships among firms' (Schoenberger, 1988:256). Hence, the call for a rejuvenation in the capital stock in conjunction with a similar transformation in labour practices, is reinforced by the matched plant analysis.

Previous results pertaining to labour flexibility illustrated significant differences between the Northern Ireland and Massachusetts firms. In terms of remuneration, the Massachusetts employees displayed a greater polarization between the unskilled and other workers, than that of Northern Ireland firms. That is, within the Massachusetts firms there was a much clearer distinction between the core and peripheral work force, with the unskilled Massachusetts workers earning a lower income relative to the other skill groupings.

With regard to the functional flexibility of employees, the Massachusetts firms illustrated a higher incidence of multi-skilling than their Northern Ireland counterparts, with this being compounded in the greater use of teams and quality control circles within the Massachusetts engineering firms. Overall, the MA employees held a significantly higher percentage of qualifications - especially 'higher' level qualifications, such as degrees, Masters degrees and PhD's - with a substantially lower percentage of employees having 'no qualifications' as compared to the Northern Ireland firms. Therefore, it can be stated that although the Massachusetts firms displayed the characteristics of a core-periphery work force to a much greater extent than the Northern Ireland firms, even these 'peripheral' workers were better qualified than their equivalent employees in Northern Ireland.

In addition to these findings, other factors such as the more positive attitude of Massachusetts employers to training their workers and the substantially higher investments in training, reinforce the perception of Massachusetts employees being more highly trained. Additionally, employees within the Massachusetts firms were characterized both by a greater elimination of lines of demarcation, and the greater interaction and interdependence of employees. The Massachusetts employees were capable of a range of tasks throughout the shop-floor, a factor which stood in sharp contrast to the rigid demarcation and task specific nature of employees within the Northern Ireland firms.

The Massachusetts employees were capable of performing a wider range of tasks throughout the shop-floor than the Northern Ireland employees, not only in terms of the aspects of production which they could perform, but also in terms of the manner in which their jobs were intensified. Thus the manifestation of the Massachusetts employees approximating the ideals of Just-In-Time to a larger extent than the Northern Ireland employees has significant implications in terms of the comparative intensity of working practices. For example the underlying characterization of Just-In-Time has been described as being; 'a highly developed form of work intensification which belies any notion of job enrichment through team-working, flexibility and job rotation, claimed by many of the proponents of JIT... but JIT aims to eliminate all costs surrounding the production process - of equal importance are the unproductive and wasteful elements of the worker's labour such as waiting time, down time and set-up times' (Turnbull, 1987:8 as quoted in Tomaney, 1990:36).

Therefore it can be concluded that the Massachusetts employees were upgrading their skills and undertaking a wider range of tasks within the firm. Conversely, within the Northern Ireland firms while there have been technological advancements as measured by the age of machinery and the presence of advanced technology, the benefits from such have not been realized due to the failure of these firms to adapt the skills of their labour forces, in line with the capital requirements.

Piore and Sabel (1984:161) emphasize that the multi-skilling of workers is a prerequisite to the implementation of Just-In-Time principles of production. Therefore, irrespective of the capital equipment within these small manufacturing firms, unless the employees have the appropriate skills base to operate an efficient and flexible system of production the labour force inadequacies will merely serve to reinforce the inadequacies of the prevailing form of production. Additionally, it is not only the skill capabilities of employees which will have to change, but also the willingness of employees to adapt to flexible forms of production. As this analysis has suggested, the introduction of flexible labour practices is directly correlated with an intensification in the utilization of labour. If the employees of these firms resist the intensification of their work practices then, irrespective of the skill levels of employees, such a system will not succeed.

It can be concluded therefore, that there is a direct and positive correlation between the educational competences and training received by employees and the extent to which the Northern Ireland and Massachusetts matched plants have adopted Just-In-Time forms of production. In establishing this correlation, an assumption is made about the competitive advantages within the Massachusetts firms. Just-In-Time is recognized not only as 'a low inventory production system but [also as] a comprehensive manufacturing style aimed at the complete elimination of unproductive activities and waste, and the maximum time-compression of productive labour performed on productive capital' (Swyngedouw, 1991:321). Therefore, the Massachusetts firms, in approaching more closely the principles of Just-In-Time production, will be comparatively more competitive than their Northern Ireland counterparts.

Notes

1 It should be noted that this analysis is based upon the assumption that each firm will seek to minimize its costs, with these including, transportation, information and financial costs associated with production. Additionally, for reasons associated with JIT, it will be advantageous if the suppliers are close to the producers should demands change, or components be lost due to faults within the system etc.

2 By asking the matched plants if they produced goods to be held as inventory, this overcame the problem whereby firms may have had a quantity of goods within the factory, which were awaiting delivery, or were at that time a result of surplus production. The extent to which the matched plants manufactured goods for stock was therefore attained.

3 This greater propensity to manufacture for stock amongst Northern Ireland engineering firms (allowing for the spatial positioning of markets) may reflect a lower technological capability both in terms of vintage and sophistication of capital, thus rendering the Northern Ireland firms unable to undertake flexible production methods.

4 It should be noted that, where there are a small number of contracting firms, this does not automatically imply that there is a high level of market vulnerability. Attempts were made to attain information relating to the percentage of output which contractual relationships accounted for. For many firms however, the sporadic nature of subcontracting orders created difficulties in estimating total subcontracting as a proportion of total sales.

5 Senker (1988:10) points out that within the developed countries, per capita calorie intake is near saturation, with it declining between 1970 and 1980 by 5 per cent. Senker concludes that 'any underlying trend of almost static population growth in the UK means that the domestic market for food has very little potential for real expansion.'

6 Estimates have suggested that 40 per cent of total food sales are attributable to only 30 firms within the UK (Beaumont, 1982:7).

7 A good example of 'relational' subcontracting in the manufacture of food products occurs between Mark's and Spencer and the principles under which they trade with their subcontractors. This is outlined in Rees, 1969.

8 This does not necessarily mean that R&D will be unplanned or unstructured. Indeed, where R&D is undertaken informally this may demand greater production planning, greater adaptability of employees throughout the workplace and greater flexibility of workers in working time.

9 Growth being defined as an overall expansion in the activities of the firm, incorporating such variables as employment, turnover, productivity, profit levels and volume of output etc.

10 It should be noted that where collaborative R&D led to product developments and innovations, contractual agreements were usually established between the firms. While this may be regarded as a rigidity, it provided a basis upon which small firms could develop their innovative capability and benefit from the expertise, marketing etc. of a larger company.

11 It should be noted that none of the small firms visited had introduced FMS, largely due to the financial constraints associated with such. Therefore the supposition of Piore and Sabel (1984) that there would be a widespread adoption of FMS within the small firm sector of the economy, is questioned.

8 Conclusions

Flexibility - rhetoric or reality?

It has been demonstrated that traditional theories of industrial location are no longer adequate explanations of regional development. Instead of the economic prosperity of regions being determined by product life cycles, and the international division of labour, regional prosperity is closely related to the inherited characteristics, the behaviour of firms, the innovative milieu and regional policy adopted within each region. It is the combination of these factors which break the dependency of regional economies upon national and international hierarchies of development and allow regions to operate independently in determining their own technological trajectory. The ability of regions to determine their own economic development and prosperity has important implications for those areas which were traditionally labelled peripheral. It follows that peripheral regions can now escape from the vicious circle of economic marginalisation and reposition themselves in the international economic order.

Flexibility is perceived as vital to regional development. Flexibility, as is most succinctly emphasized by the Regulationists school, extends beyond the confines of the firm and pervades social and economic institutions alike. Moulaert and Swyngedouw (1989) emphasise that flexibility embraces not only the production process and the labour market, but also the manner by which production is geographically organized, the role of government intervention and support and the ideology of society. While acknowledging the scope of flexibility the investigation presented within this book has concentrated upon the conditions which directly impinge upon the firm. In particular, attention has been focused on the small firm sector, as this is the area towards which substantial government assistance has been targeted.

This research has examined a sample of small engineering and food processing firms in an attempt to ascertain the myth or the reality of corporate flexibility. In determining the nature and presence of corporate flexibility the extent to which this impacts upon growth has also been measured. As the results have demonstrated, it is unequivocal that the Massachusetts firms had adopted flexibility to a greater extent than their Northern Ireland counterparts. In addition, higher functional flexibility and flexibility in production within the Massachusetts plants were associated with higher levels of small firm growth as measured by sales per employee and change in turnover from 1988 to 1993.

Higher educational and vocational qualifications amongst the Massachusetts employees combined with significantly larger investments in training, have facilitated greater functional flexibility than that found within Northern Ireland. While employee competencies are essential to the introduction of functional flexibility it is the awareness of the need to organize strategically production in a flexible, responsive manner that determines the extent to which functional flexibility becomes an integral part of the production process. Through functional flexibility the organization of labour is restructured in a more efficient and productive manner. This organizational structure has direct implications upon inventory control, quality control and the innovative capacity of the firm. As has been demonstrated, it is these changes which have important implications for corporate performance.

Flexibility and growth

The model of the flexible firm (Atkinson, 1984:28) suggests that firms will search for three types of flexibility, including numerical, functional and financial flexibility. This flexibility can also be described as adaptability as it describes the ability of firms to respond to changing market conditions and technological adjustments. The Flexible Specialization approach proposes that the market and technological conditions which characterized Fordism have been transformed, with flexible firms now responding quickly 'to sudden changes in costs, market opportunities and/or new technologies, through adopting flexible, multi-purpose equipment and creating a flexible, re-integrated and co-operative work force free of the shackles of rigid job specifications, narrow job-centered orientations and excessive regulation and control' (Wood, 1989:11).

It would be anticipated therefore that where a firm adopts a policy of flexibility within both labour and production, then this will create benefits in two ways. Firstly, the firm will be better equipped to be *reactive* to changing demands and therefore gain competitive advantages over its less flexible competitors. Secondly, flexibility will establish the conditions for the firm to be *pro-active* in minimising costs through the efficient organization of production and the performance of innovative activities which are directly associated with growth (Schumpeter, 1961; Coyne and Wright, 1975; Adams,

1982). Within the context of the Northern Ireland and Massachusetts matched plant analysis it would be expected therefore that higher growth would be associated with greater flexibility.

The matched plant analysis of small engineering and food processing firms demonstrated that firms within both regions had experienced turnover and employment growth in the period 1988 to 1993. Yet, the extent to which firms had experienced growth was not evenly distributed for the samples. Both regional and sectoral disparities were found in terms of turnover and employment growth. With the exception of mechanical engineering, the percentage change in turnover for the Massachusetts engineering firms greatly exceeded that for the Northern Ireland firms. Further, turnover growth within the food processing matched plants, between 1988 and 1993, was accompanied by a substantial growth in employment although the ratio of turnover to employment growth varied between the regions. Within the Northern Ireland food processing firms a 6.15 per cent increase in turnover was coupled with a 13.5 per cent increase in employment. Conversely, a 48.6 per cent increase in real turnover within the Massachusetts food processing firms was accompanied by a 27.5 per cent increase in employment. Even within a traditional industrial sector such as food processing, the Massachusetts firms were more efficient in their utilisation of labour and had increased their output per employee over the 1988 to 1993 period by a level significantly above that of their Northern Ireland counterparts. Based upon turnover growth, it can be concluded that the Massachusetts engineering and food processing firms had grown significantly faster than the Northern Ireland matched plants over the 1988 to 1993 period.

Having established this disparity in turnover growth the remainder of this section recapitulates the main findings concerning the adoption of flexibility practices within the firms. The manner by which the Massachusetts firms outpaced the Northern Ireland firms in the levels of turnover and sales per employee growth from 1988 to 1993 is explained in terms of the presence of corporate flexibility. It is evident from this that numerical flexibility was not widely adopted by the engineering firms within either Northern Ireland or Massachusetts, yet, was used more within the food processing than in the engineering firms. The evidence suggests that numerical flexibility did not contribute to growth within the engineering matched plants, and indeed it is possible that where engineering firms adopted numerical flexibility this may have led to increased costs, lower productivity and increased productive inefficiencies.

Functional flexibility was significantly more apparent within both the Massachusetts engineering and food processing matched plants, as evident in the higher qualifications and training of all employees; the greater elimination of lines of job demarcations; the formation of work teams and quality circles; and the restructuring of the managerial-labour hierarchy of control. The Massachusetts matched plants were more capable of adjusting and deploying their employees throughout the workplace and this was reflected in their comparative advantage over the Northern Ireland matched plants in both performance and growth.

The final indicator of flexibility within the firm related to flexibility within the production process. Again significant disparities were found between the Northern Ireland and Massachusetts matched plants with the Massachusetts firms being more responsive to changing market demands; they had more developed subcontracting relationships and inter-firm linkages which contributed to a higher innovative capacity; incorporated quality control more efficiently within the production process which led to lower reject rates of production and experienced comparatively fewer capital breakdowns than their Northern Ireland counterparts.

Numerical flexibility and growth

There was an industrial bias in the utilization of numerical flexibility, with the food processing matched plants adopting numerical flexibility to a greater extent than the engineering firms, within both Northern Ireland and Massachusetts. Food processing, being relatively low skilled demands a low level of education, training and adaptability from its employees. This therefore enables firms to 'hire and fire' as a direct response to changing market conditions. In contrast, within engineering the majority of tasks are semi-skilled or skilled and therefore a higher level of competence is demanded from engineering employees, with this aptitude being acquired through education and training. Part-time and temporary workers, in acting as a cost reduction mechanism in times of demand uncertainty and vulnerability, were used more extensively where tasks were low-skilled, training requirements for new recruits were low and there was a ready supply of suitable workers in the external labour market.

While it is proposed that the incidence of collective power will restrict the introduction of numerical flexibility within firms (Farnham, 1990), this was not supported within the matched plant analysis. Instead, for the engineering firms, there was a higher union presence within the Massachusetts firms, yet these firms also had a slightly higher incidence of numerically flexible workers. Further, where unions are present it would be anticipated that there would be a smaller remuneration gap between skilled and unskilled workers, with the overall levels of pay being higher than within non-unionized firms (Metcalf, 1988). Yet, this was not supported within the matched plant analysis. While the presence of unions was higher within the Massachusetts sample this did not correspond to higher overall wages and instead a greater peripheralization as based upon pay levels was found between unskilled (in the engineering sample), and unskilled and semi-skilled (in the food processing sample) relative to the 'core' workers in the Massachusetts firms.

No significant correlation was found in the extent to which firms had implemented numerical flexibility and the growth levels of the firms between 1988 and 1993. This does not imply that the adoption of an organizational strategy based upon numerical flexibility will not affect *profit performance*, yet, within the matched plant analysis numerical flexibility and *sales growth* were not interdependent.

Functional flexibility and growth

The micro-dual labour market model (Atkinson, 1985) of the flexible firm proposes that core workers will be functionally flexible while the peripheral workers will be numerically flexible. The core workers are central to the firm's operations and are differentiated from the peripheral workers in terms of pay, status and job security. The higher job security is due to their skill levels and the limited availability of similar workers in the external labour market. While functional flexibility enables firms to respond to demand uncertainty or variability, functional flexibility is more closely linked to changes in the operational environment of the firm. Of central importance in the attainment of functional flexibility is the training of workers to a point where they are multi-skilled. While changing demand and production forces within firms, and the introduction of new technology provides the impetus for creating a core work force which is functionally flexible, functional flexibility can only be attained through the training/re-training and multi-skilling of employees. A number of other studies have identified a significant and positive link between training and competitiveness (Office of Technology Assessment, 1990; NIESR, 1991; Prais and Wagner, 1988; Roper and Hofmann, 1993). In differentiating between skills and growth, and skills as the facilitator of functional flexibility, these associations are examined within the context of the matched plant findings.

Employee skills in stimulating growth

The relationship between employee skills, productivity and competitiveness has been documented widely (Office of Technology Assessment, 1990; NIESR, 1991; Prais and Wagner, 1988; Roper and Hofmann, 1993). While the identification of productivity and competitiveness gaps are accurately ascribed to comparatively lower skills, the manner by which skills impinge upon the production process is less well understood. For example, it has been recommended that; 'If Northern Ireland firms are to become more internationally competitive they will need management and work forces that are as skilled as their international competitors' (Roper and Hofmann, 1993:68).

Yet, as the Northern Ireland, Massachusetts matched plant analysis has demonstrated, while skill enhancing and multi-skilling are directly associated with improved performance, the multi-skilling of workers merely creates conditions by which a competitive advantage can be attained, through the flexible deployment of employees.

The Northern Ireland and Massachusetts matched plant analysis has demonstrated that while a significant gap existed in the skill levels of employees, with this being further accentuated by a higher investment in the training of Massachusetts employees, this alone did not provide an explanation for the disparities in performance and growth between the regions. The analysis has shown that skill levels merely provided a foundation upon which economies of scope can be realized. Within the Massachusetts firms it was not the skills of the employees *per se* which were directly affecting performance, but rather the way in which the skills permitted efficiencies in the production process to be made.

Therefore, the introduction of flexibility within the production process can only be facilitated through the pre-condition of highly skilled workers. These skills then enable the firms to introduce functional flexibility, which is recognized as 'the ability of firms to adjust and deploy the skills of their employees to match the changing tasks required by a changing workload, production methods or technology' (Pinch, Mason and Witt 1991:208).

Employee skills in facilitating functional flexibility

The matched plant analysis has demonstrated that the ability of firms to implement functional flexibility is dependent upon the skill levels of the work force. Overall, the Massachusetts employees had attained higher formal educational and vocational qualifications than the Northern Ireland employees. This disparity in qualifications was greater for the engineering, than the food processing matched plants. Amongst the food processing firms although the overall qualificational attainment was slightly higher within Massachusetts, the disparity was not sufficiently large to state that the better performance and higher growth levels within the Massachusetts firms were a direct result of employees' qualifications. Conversely, amongst the engineering firms there was a large difference between the regions with regard to the percentage of the work force without formal qualifications. Over half of the Northern Ireland employees had no qualifications compared to approximately a quarter of the Massachusetts employees. Additionally, a higher percentage of the Massachusetts workers had completed certificate courses, Associate degrees, Baccalaureate degrees, Masters and PhDs than within Northern Ireland. In examining the extent to which higher skills, and subsequently, functional flexibility were confined to the core workers (Atkinson, 1984), analysis of skill levels and functional flexibility was made for managers, supervisors and shop floor workers separately.

Managers

Managerial qualifications have been identified as being the most significant in determining the strategic direction of the firm. For example, while the managers of engineering firms within Northern Ireland surpassed the Massachusetts managers in the incidence of BTEC/Associate degrees, and were comparable to the Massachusetts managers in the attainment of Baccalaureate degrees, the Northern Ireland managers considerably trailed behind the Massachusetts managers in the attainment of Masters and PhD degrees. A number of other matched plant studies (Hitchens, Wagner and Birnie, 1990; Roper and Hofmann, 1993; Roper, 1994) have illustrated similar results to that found within the Northern Ireland, Massachusetts study, for the qualification levels of both the work forces as a whole as well as the managers, and these studies have concluded that the higher level of managerial skills has enabled the production systems to operate more effectively through the relatively easy adoption of new product developments (Roper and Hofmann, 1993). It can be concluded that the higher qualification level of managerial workers within the Massachusetts firms will culminate in the attainment of competitive advantages for the firms concerned, through the more efficient running of the production process.

Other research (DED, 1987) such as the 'Pathfinder' report highlighted the importance of developing; 'quality, design, product innovation, technology transfer and R&D' (Harris, 1993:82), within Northern Ireland firms. However, without the necessary managerial expertise and training it is unlikely that these functions will be realized. Indeed studies such as those by Hitchens and O'Farrell, and Hitchens and Birnie, (1989, 1990) along with those of Mason, van Ark and Wagner (1994) have consistently illustrated that the design and quality of products within Northern Ireland and British plants have been of an inferior standard to that of matched plants within Germany, the Netherlands and France. Additionally, each of these studies also found that the Northern Ireland matched plants had lower levels of qualifications compared to other regions. Roper (1994:8) concludes that the higher skill levels as found within the comparators of the Matched Plant studies, have culminated in the competitive advantage of firms through 'increasing the firms' commitment to the development of new products, by easing the adoption of new products and processes, by increasing production efficiency by reducing labour costs, and by facilitating high quality training'.

In summary, the Northern Ireland, Massachusetts matched plant analysis has demonstrated that there was a direct and positive relationship between the skill levels of managers and the efficiency of the firms as measured by the inventory levels; the reject rates of production; the introduction and development of new products and processes; the performance of R&D activities; and the rate of turnover growth.

Supervisors

The expected role of supervisors within the Massachusetts firms was characterized by the identification of ways in which to increase efficiency, with training accentuating this role by increasing their ability to develop new products and processes. In contrast, the supervisors within the Northern Ireland firms were perceived as trouble-shooters in overcoming technical difficulties within the production process. As supervisors within the Northern Ireland firms received a limited amount of formal training, this restricted their roles and hindered the development of more analytical and innovative skills. Additionally, restructuring of the role of supervisors within the Massachusetts engineering firms was apparent with a 19 per cent reduction in the number of supervisors from 1988 to 1993. This reduction in the number of middle-managers was done without affecting the efficiency of production. Rather, the reduction in the number of supervisors was facilitated through an overall improvement in the work force. For example, shop floor employees were trained to take more responsibility for their own work with this being evident in the development of work teams to monitor the quality of production and the undertaking of continual preventative maintenance of machinery.

Shop-floor workers

The matched plant analysis demonstrated that the Massachusetts shop floor workers displayed comparatively higher levels of qualifications, as compared to the Northern Ireland shop floor employees. This disparity in the levels of qualifications was greater within the engineering sector than the food processing sector - where both regions employed a largely unskilled work force in the performance of relatively low-skilled jobs. Roper and Hofmann (1993) also discovered that cumulatively, the shop-floor employees within the Northern Ireland firms had significantly lower skill levels than those within the German firms. Hence, they concluded that within the clothing sector, as a direct response to the skill levels of employees the German shop-floor workers, 'could perform five or more operations compared to only 5-10 per cent in each Northern Ireland firm' (Roper and Hofmann, 1993:59).[1] Similarly, the higher qualification levels amongst Massachusetts shop-floor employees and the multi-skilling of these workers through higher investments in training contributed to the superior performance of the Massachusetts firms, through the reorganization of production. Hence, the higher skill levels of shop floor workers coupled with those of supervisory, technician and managerial staff, together created the necessary preconditions upon which the firm could flexibly deploy its labour in response to changing market demands and technological adjustments.

Amongst the Northern Ireland engineering firms, where the market is largely protected and characterized by a constrained external and internal labour market, this leads to firms being less willing to provide workers with the financial and occupational assistance to further their skill and qualification levels as similarly low levels of training within other firms will not entice workers away from their present employers.

As the Northern Ireland market is sheltered and dominated by 'experience' goods (Clulow and Teague, 1993), with approximately two-thirds of the output of these firms remaining within the province (Harrison and Hart, 1993; NIERC, 1993) a lower quality of products (Hitchens, Wagner and Birnie, 1990), and an inferior level of design and production has resulted (Mason, van Ark and Wagner, 1994). There is therefore, not an immediate need to raise the standards of workers as their skill levels, albeit low, are adequate in the production of goods for the home market.

These factors merely reflect the sub-optimal level of skills present within the Northern Ireland firms. Yet, what is of greater concern is the disparity between the two regions in the amount of training which the employees received. Accounting for the gap in educational and vocational qualifications between the Northern Ireland and Massachusetts firms, it would have been anticipated that the Northern Ireland matched plants would have invested a substantially larger amount of money in the training of their employees. This was not however found. Although the entry-skills threshold within the food processing sector is not as great as within the engineering industry, yet in both instances job-specific skills are required by the workers. Nevertheless, there were major qualitative and quantitative differences between the engineering and food processing firms, and between the Northern Ireland and Massachusetts matched plants in the extent to which employee training was provided.

The same feature of lower skill levels along with a deficient training system were also found by Roper (1994) in his matched plant analysis of firms within Northern Ireland, Germany and Italy. Again it was found that the lower skill and qualification levels of Northern Ireland employees, was reinforced by a failure of the Northern Ireland firms to invest in the training of their work forces. Within the Northern Ireland, Massachusetts study, it was demonstrated that the Northern Ireland firms invested significantly less in the training of all their employees as compared to the Massachusetts firms. Although the Massachusetts employees surpassed the Northern Ireland employed in the attainment of educational and vocational qualifications, the Massachusetts firms still invested significantly more in the training of their employees. Indeed, values for the per capita expenditure on training reinforce the regional disparities in employee training, with the Massachusetts firms attempting to attain coordinated development and expansion through an overall upgrading of employee skills.

While the flexible firm model perceives functional flexibility to be found amongst core workers, the Massachusetts engineering firms through training, skill enhancing and multi-skilling were also able to apply functional flexibility to their shop floor workers. Where unskilled and semi-skilled employees displayed the capacity to be numerate, solve problems, and act in an intuitive and intelligent manner upon the shop floor, then this had significant effects in terms of the efficient running of the firm, an increase in productivity, a reduction in down-time due to machinery break-downs and changes and the overall improvement in the quality of production as workers become more apt at identifying problems at an early stage as well as identifying areas for improvement within the system.

Flexibility in the production process and growth

Fordist principles of production were centered around the technical division of labour, which facilitated the continuous production of standardized products on a large scale for mass markets. While the Flexible Specializationists and Regulationists disagree on the causes and determinacy of the crisis of Fordism, both point to the presence of internal structural problems within mass production. These structural problems created a mis-match between productive capacity and demand, whereby rigid and dedicated technology was incapable of adopting to differentiated and fragmented markets[2]. Both theoretical bodies point to the emergence of new, advanced production technology as providing a means of overcoming the technical problems within Fordism (Meegan, 1988). New advanced technology would permit the reorganization of production away from large scale mass produced goods to small batches of customized goods without reducing operating efficiency within the firm. This would therefore allow small firms to gain a competitive advantage, through economies of scope as opposed to the traditional competitive advantage of large firms as based solely upon economies of scale.

The introduction of flexible technologies would more closely aligned the productive capacity of the firm to the demands of differentiated and fragmented markets. While new advanced technology is recognized as being vital to overcoming the crisis of Fordism, it merely facilitates the reorganization of production in a way which enables firms to switch quickly between different products and processes in response to changing demands. Changing market conditions also have implications for the manner by which firms interact with each other (Scott and Storper, 1988; Storper and Walker, 1989). The strategic organization of production will therefore adopt flexible production strategies which have been modelled on Japanese forms of subcontracting, namely Just-

In-Time systems of production. While JIT systems have been defined as 'low inventory systems' examination of the Japanese experience demonstrate that 'if anything, reduced inventories have been a consequence of sharply reduced changeover times and lead times, source inspection methods, and self-stopping machines' (Best, 1990:143). Indeed, flexible production depends 'first, upon the integration of planning and doing on the shop floor and associated redefinition of the roles of both management and labor, and second upon consultative supplier relations' (ibid). Therefore, while labour flexibility provides firms with the *ability* to deploy workers flexibly throughout the firm, this merely allows production to be organized in a flexible and efficient manner, which facilitates the attainment of economies of scope and improves the overall operating efficiency within the firm. Functional flexibility while allowing workers to be transferred throughout the production process and across the traditional technical division of labour, can only be implemented where production is organized flexibly. The matched plant analysis examined the extent to which firms had introduced flexibility within their production processes. As is now summarized, significant disparities were apparent between the Northern Ireland and Massachusetts firms, in inventory control and the incidence and characteristics of subcontracting relationships, with little difference being found in the capital investments within the firms.

Inventory control and the elimination of buffer stocks

Examination of the proportion of firms which manufactured goods for stock demonstrated that 44 per cent of the mechanical engineering firms in Massachusetts and 66 per cent within Northern Ireland, manufactured goods for stock. This regional disparity widened with regards the electrical and electronics matched plants, with 33 per cent of the Massachusetts firms, as compared to 66 per cent of the Northern Ireland firms manufacturing for stock.

For firms within the metal goods and food processing sectors, there was comparability in the proportion of firms which manufactured goods for stock. Within the metal goods sector, firms were manufacturing a large number of product variations and therefore this constrained the feasibility of manufacturing for stock. Amongst the food processing firms, product perishability and the relative ease with which operatives could be transferred between assembly lines reduced the necessity for buffer stocks within both regions. The significantly lower proportion of mechanical engineering and electrical and electronics firms within the Massachusetts sample which manufactured goods for stock is closely associated with higher skill levels, a reduction in job-demarcation and the increased interdependence of workers in the form of team-working and quality control circles. Therefore, the results suggest that there is a positive association between the ability of a firm to introduced functional flexibility (as measured by the educational and vocational qualifications, and the extent to which employee skills are upgraded), the degree to which labour is flexibly deployed throughout the firm (as measured by the elimination of both lines of demarcation and direct control and supervision) and the quantity of goods which are held as buffer stocks.

Extent and nature of subcontracting relationships

Results from the matched plant analysis illustrated that while subcontracting relationships were significant within both regions a larger percentage of the Massachusetts matched plants engaged in subcontracting activity. While large and small firms have always interacted to some extent, the nature of these relationships has changed over time. Aydalot (1988:171) states that, 'Over the past century industrial complexes have gradually been developed. At the same time, inter-industry relationships have also been created so that an industrial milieu has been generated by combining inter-enterprise relationships, the most typical example of which is subcontracting.'

While the incidence of subcontracting relationships has increased significantly since the 1970s (Aydalot, 1988), in many cases the form which these relationships have taken has remained unchanged. For example, while subcontracting was important for both the Northern Ireland and Massachusetts matched plants, the nature of these relationships varied both regionally and sectorally. For most of the Northern Ireland matched plants which were engaged in subcontracting, this took the form of traditional, low interference manufacturing. In many cases the small firms were subcontracting for the same firms as they had five, or ten years previously, with the firms often being overly dependent upon a small number of buyers. For the contracting firms, the establishment of long-term subcontracting relationships with these smaller suppliers provided a stable market base upon which they could expand their market base while externalising the risks of production. The high level of market dependency upon the contracting firms often resulted in an acute vulnerability, which in the eventuality of market failure or a down-swing in the fortunes of the contracting company could culminate in the closure of the subcontractor.

Within each of the industrial sectors examined, a higher proportion of Massachusetts firms were engaged in subcontracting. Yet, on average, the share of total output which subcontracting relationships accounted for was less than that for the Northern Ireland firms. In addition, the average Massachusetts firm undertook subcontracting with a greater number of contractors than the Northern Ireland firms, thus reducing their overall market dependency and operating vulnerability. Yet of greater significance with respect to the nature of subcontracting relationships was the format which they took, especially amongst the electrical and electronics firms. Within Massachusetts this sector is closely linked to the high-technology agglomeration within Boston, and this had significant effects upon the subcontracting relationships which were found. All of the electrical and electronics matched plants within Massachusetts were involved in subcontracting, with the nature of this departing radically from the traditional, classical format as found within the Northern Ireland matched plants. Subcontracting within these firms was characterized by strict codes of conduct and the establishment of relational or strategic alliances (Clulow and Teague, 1993). These strategic alliances were characterized by the co-ordination of research between the contractors and subcontractors, whereby the R&D laboratories were often made accessible to the technical staff of the small firms. While the small firms were constrained in obtaining exclusive patented rights to the development of new products, the financial benefits of these developments could not have been realized in the absence of collaborative arrangements.

While virtually any small firm may engage in subcontracting relationships with larger enterprises, the changing demands of contractors is radically altering the underlying demands being placed upon the small manufacturing firm. For example, contracting firms are increasingly demanding shorter lead times and the contraction of demand and delivery times. This combined with the demand for Total Quality Management, and the production of high quality products have all affected the nature of traditional subcontracting relationships. If small firms can strategically organize their production processes in line with changing demands, then they will maintain and increase their market share. If however they fail with respect to the introduction of technologically advanced machinery, the multi-skilling of their workers, or the adoption of flexible forms of production including JIT and TQM then their competitiveness position will be erased and market share will decline.

Within the Massachusetts region, and particularly within the electrical and electronics sector, the complementarity between large and small firms is manifest less through the externalization of risks and the exploitation of the small firms (although this undoubtedly occurs for many firms), but rather is characterized more through the fostering of strategic alliances. If the Northern Ireland firms are to attain the economies which are possible through the competition and co-operation of large and small firms then a number of factors will have to be altered. For example, the region will have to advance the overall skill levels of its work force to provide the occupational base upon which larger firms and entrepreneurs can build businesses which are functionally flexible. Secondly, through the reduction in the 'skills gap' (Roper, 1994) of Northern Ireland firms relative to their competitors and the subsequent propagation of high-value added small firms, inward investors could be encouraged to locate their higher order levels of production within the region as opposed to merely their low value-added, low skilled branch plants. This would contribute significantly to GDP particularly if advanced engineering as opposed to food processing and clothing firms located within the region.

Finally, the accessibility for small firms of an adequate supply of venture capital should be developed. In the absence of banking and other financial institutions which are prepared to fulfill this function, and the initial absence of large firms with the resources to finance projects within small firms, the government should intervene and provide this function in a professional capacity, with the strict monitoring of such projects and the repayment conditions which are characteristic of the private sector.

If relational and strategic alliances are to be developed more fully within the Northern Ireland firms, it is essential that small firms create the conditions of a flexible and functionally adaptable work force, which behave symbiotically with the flexible organization of production, both within the firm and through inter-firm linkages.

The matched plant findings have demonstrated that the incidence of subcontracting relationships did not have a direct positive or negative effect upon the growth levels of the firms. Rather, the *nature* of the subcontracting relationships bore a substantial influence upon the resultant growth levels. For example where the subcontracting relationships were characterized by the provision of venture capital, assistance with the performance of R&D and access to highly skilled and specialized personnel - both within the contracting companies and the research institutions - this had a positive impact upon turnover growth within the matched plants.

Age and presence of technologically advanced equipment and growth

The Flexible Specializationists (Piore and Sable, 1984) believe that the crisis of Fordism, as found within the Third Italy, has been overcome through the 'flexible use of increasingly productive and widely applicable technology... (flexible production technology) had to be flexible in both a narrow and broad sense. It had to permit quick, inexpensive shifts from one product to another within a family of goods, and it had to permit a constant expansion in the range of materials worked and operations performed, in order to facilitate the transition from one whole family of products to another' (Piore and Sable, 1984:30). As the industrial hegemony of large firms was challenged with the ability of small firms to gain competitive advantage based upon economies of scope, a positive relationship emerged between the adoption of technologically advanced machinery, increased market share and the profit performance of small companies.

Evidence from the Matched-Plants survey illustrated that the Northern Ireland firms had a higher proportion of capital equipment under five years[3] of age. Further analysis demonstrated that the Massachusetts firms had a higher proportion of technologically advanced machinery, as characterized by computer aided design and numerically controlled machine tools, however this was not sufficiently large enough to explain the superior performance of the Massachusetts firms.

Other matched plant studies have also found similar results regarding the composition of capital equipment within Northern Ireland firms. Hitchens and O'Farrell (1987, 1988a, 1988b) in their matched plant analysis of firms in Northern Ireland, Scotland, Wales, Southern England, Mid-West Ireland and Dublin also discovered that 'a comparison of machinery used indicated substantially more modern and more computerized equipment in Northern Ireland' (Enderwick et al., 1990:140). Similarly in a Northern Ireland, Germany matched plant analysis it was found that while the overall age of capital equipment was similar between the matched plants, more computerized design facilities were found within the German companies (Roper and Hofmann, 1993).

Flexible technologies are heralded as providing small firms with competitive advantages based upon economies of scope, the capturing of niche markets and higher profit performance due to the efficiencies in production (Piore and Sable, 1984). Simultaneously the incorporation of flexible technologies within the firm is believed to correlate to the upgrading of employee skills. Results from the Northern Ireland, Massachusetts matched plant analysis fail to support these claims. Instead it is found that there is no significant correlation between the age and technological sophistication of capital equipment and the functional flexibility or the growth levels within the firms.

Numerical, functional and production flexibility - Mutually exclusive concepts?

The existence of an interdependent relationship can be identified between the various components of flexibility. Within this relationship it is found that numerical flexibility may occur extensively within an industrial sector which demands a low level of competence from the employees, in terms of the tasks which they perform. Such is manifest within the food processing sector, where it was apparent that a large percentage of the work-force, within both Northern Ireland and Massachusetts had no formal qualifications, and such employees received a very low level of investment with respect to the furtherance of their skill levels. In such circumstances therefore, the objective of numerical flexibility may be implemented within the policy remit of the small firm with relative ease. The relative difficulty of introducing such as an employment policy within the small engineering firm sector is however increased, with numerical flexibility being closely related to other factors of production.

Within the engineering industry, the ability of firms to adopt a policy of numerical flexibility will depend upon the presence of suitably qualified employees in the external labour market. As most employees within the engineering industry must be at least semi-skilled, with the introduction of new technology demanding that they upgrade their skills to incorporate computer systems knowledge, this will reduce their fluidity within the external labour market as compared to un-skilled assembly line workers within the food processing sector.

The results suggest that there is an inverse relationship between the presence of numerical flexibility and functional flexibility. In general, it was found that the ability of firms to introduce functional flexibility was associated with higher skill levels and employee training. Where investments were made in training, firms were therefore less likely to implement numerical flexibility.

Significant disparities were found between the Northern Ireland and Massachusetts engineering firms in the educational and vocational qualifications of all employees. While the regional disparity in qualifications was reduced slightly within the food processing matched plants, again the Massachusetts employees had a higher incidence of qualifications than their Northern Ireland counterparts. The skill levels within the engineering sector were higher for all sections of the work force, from un-skilled shop-floor workers to technicians and managers. Additionally, each of these categories of workers received a higher level of investment in their training than the Northern Ireland employees. Similar results were also found in a matched plant analysis of Northern Ireland firms with German and Italian firms. This study also found 'significant skill differences...which were seen to have a significant impact on firms' production efficiency and innovative capability. That is to say, a firms work force and managerial skill levels not only shape how competitive a firm is likely to be in the short-term, but also its ability to respond to changing market conditions' (Roper, 1994:10).

The higher skill capacity of the Massachusetts firms were positively correlated with a number of other characteristics within the firms. For example, within the Massachusetts engineering matched plants higher managerial skills led to the greater performance of R&D activity, which is positively correlated with a higher rate of product innovations and therefore performance (Roper et al., 1996). Further, the presence of a positive association between the skill levels of the employees, the degree to which functional flexibility was being implemented within the firm and the growth of the firm suggests that production was organized in a more flexible and efficient manner within the Massachusetts firms.

There is therefore a further link in the flexibility process between the skill levels of the workers - and the subsequent functional flexibility which this facilitates - and the flexibility of production as comprising such principles as Just-In-Time manufacturing, the implementation of quality control circles and the adoption of a framework wherein Total Quality Management can be operationalized. It is *only* through the attainment of these later aspects of production that a firm may successfully realize the competitive advantages associated with the incorporation of both numerical and functional flexibility. Hence while the advancement of skills and the educational capabilities of employees is essential, it does not provide the Holy Grail to the search for improved competitiveness. Rather, the skill and educational levels along with the training performed on employees are merely pre-conditions for the incorporation of functional flexibility, with the functional flexibility of the work force representing an essential ingredient in the adoption of flexibility within the production process, and vice versa.

The evidence for flexible specialization and flexible accumulation

For the Flexible Specializationists, 'flexibility refers to labour market and labour process restructuring, to increased versatility in design and greater adaptability of new technology in production. Specialization refers to niche or custom marketing, the apparent end of Fordism, mass production and mass standardization. Hence the concept unites changes in production and consumption' (Smith, 1991:139). For the Regulationists, the nature of flexibility is less well defined. While the Flexible Specializationists perceive the new flexibility as having arisen in response to changes in demand and technology, the Regulationists attribute the capitalist crisis of the late 20th Century to the productive limits of Fordism and rising social unrest within the Fordist era.

The Flexible Specializationists see the crisis of Fordism as being overcome through the introduction of new flexible technology. This allows small firms to gain competitive advantages based upon economies of scope which are attainable through the ability of machinery to make subtle alterations in the composition of products and manufacture a diversified product range from a smaller overall capital stock. For the Regulationists however, changes within production must also be accompanied by changes in the economic, political and social environments to a point where synergy is found between technology, labour, consumption patterns and the social institutions (Smith, 1991). Continuity is found between the Flexible Specializationists and the Regulationists in the centrality placed upon changes within production in overcoming the crisis of Fordism (Piore and Sable, 1984; Sabel, 1982; Aglietta, 1979), with the emergence of flexible technology also being identified by the Regulationists as a means of overcoming the productive limits of Fordism (Palloix, 1976). The neo-Fordist accumulation system (as identified by the Regulationists) will therefore impinge upon 'the organization of production and the workers' relationship to the means of production; the speed of capital circulation and valorization (which influences the development of management principles); the distribution of value (which allows for the reproduction and development of different social classes); the composition of demand; and the manner of articulation with those non-capitalist forms which hold an essential place for an economic formation, such as the household' (Boyer, 1990, as quoted in Peck and Tickell, 1992:5).

Both theoretical perspectives agree that the Fordist principles of production have reached their limit and that major changes are occurring within the firm as characterized by flexibility. Yet, discontinuities are apparent in the perceived nature of these changes[4]. For this reason the extent to which the matched plant results conformed with the Flexible Specialization perspective or the Regulationist perspective, are briefly summarized separately.

The evidence for flexible specialization within the matched-plant analysis

While it is acknowledged that the economic situation within areas such as the Third Italy may approach that of Flexible Specialization this research has served to accentuate the belief that Flexible Specialization is an abstract concept with little applicability to the environments of the matched plants.

The flexible specialization theory proposes that there will be a widespread adoption of advanced flexible technology in an attempt to benefit from economies of scope and serve the needs of the market, through capturing niche markets and demonstrating the ability to transfer production quickly from either one product to another, or through the production of slightly differentiated products. Evidence from this study amongst others (Morris, 1988) has illustrated that the actual adoption of flexible manufacturing systems (FMS) within the small firm sector has been virtually non-existent, largely as a result of the exceptionally high cost of introduction as well as the fact that for most small firms, despite the acclaimed flexibility of such a system, in reality it leads to 'flexible rigidities' (Dore, 1986), in both the form of production and the utilization of labour. For example, if a small firm made the high capital investment in the purchase of a FMS then it would be confined to the market area within which the product was found. Should the prevailing market change in response to altering consumer tastes and fashions or technological innovations on an industry level, then the small firm would be paralysed as it could not financially afford the change to an alternative market and the burden of redundant machinery might render the firm insolvent. Therefore, while the Flexible Specialization theorists postulate that small firms will no longer be marginalized within the new economic environment due to the potential to obtain economies of scope, this has not been found in reality. Instead, the high capital investments which such a system demands has affected negatively the capacity of small firms to introduce a system of FMS.

Although FMS were not found within the matched plants, a number of the firms used technologically advanced machinery in the form of numerically controlled equipment. These computer controlled machines were mainly found within the design (CAD), computer numerically controlled machine tools and basic robotics functions of the firms. In addition, within the mechanical engineering plants flexible technology had permitted greater flexibility of production in the form of differentiated products which had increased the market share for these firms. However, it should be concluded that while there were some signs of flexible technology within the firms this was not on the scale of that proposed within flexible specialization. As the adoption of technologically advanced machinery was similar for both the Northern Ireland and Massachusetts matched plants, this fails to explain the superior growth performance of the Massachusetts firms relative to their Northern Ireland counterparts.

The second aspect within the firms which failed to support the Flexible Specialization theory, was that relating to economies of scale. The theory proposes that flexible technology will enable small firms to challenge the traditional hegemony of large firms through economies of scope. Yet, Allen and Massey (1989) illustrate, that the flexible manufacturing systems are capable of shifting the scale of production without significantly altering the costs of production, a factor which leads to the attainment of economies of scope. Allen and Massey also argue that these economies '*augment*' the economies of scale which the manufacture of standardized goods offer, in long production runs. Additionally, within the matched plants in both regions when the firms were asked to identify their perceived advantages over competitor firms, they frequently highlighted the price of their products as opposed to the aspects as characterized by Flexible Specialization. Within the Northern Ireland firms, when discussing the attitudes of the managers towards BS5750, a number of managers stated that ultimately, it was price considerations as opposed to quality, timeliness etc. which would determine the share of the market. While flexible technology can offer economies of scope to small firms, in the absence of economies of scale which lead to a reduction in costs of production and subsequently the price to the consumer, the viability of small firms in the market place will be confined to niche markets.

The third aspect of flexible specialization which was not supported within the matched plants relates to enhanced functional flexibility within the firms. While Piore and Sabel (1984) propose that the incorporation of flexible technology will reduce the number of job demarcations within the workplace and lead to the multi-skilling of workers, this view has been challenged widely. Others (Shaiken et al., 1986) have proposed that the introduction of flexible technology may not lead to a significant change in the skills demanded from workers. Indeed, in many cases traditional semi-skilled jobs may merely include extra maintenance and quality control functions. Hudson states that, 'the introduction of new conditions of work is associated less with the introduction of new technology than with the intensification of work with existing technology' (Hudson, 1989:8). Within the matched plant study similar findings to Hudson's were found. If the incidence of technologically advanced machinery within the firm is directly related to the reduction in job demarcations then it would be expected that due to comparability in the incidence of advanced machinery between the regions a similar pattern of job-sharing and multi-skilling would also be evident. This however was not found, but instead significantly higher levels of functional flexibility were manifest within the Massachusetts firms as compared to the Northern Ireland matched plants. This higher functional flexibility within Massachusetts was due to the higher educational and skill levels of employees coupled with greater per capita investments in training. While the presence of technologically advanced machinery may have been important in stimulating high levels of investment in training, it was found that the functional flexibility of the workers was a function of the form of production adopted within the firms. For example, while production within the MA firms was characterized by flexibility in terms of JIT, reduction of buffer stocks, the advancement of subcontracting relationships and strategic alliances etc. this was not manifest within the Northern Ireland firms. Thus, it is not the presence of flexible technology *per se* which determines the extent of functional flexibility within the firms, but instead it is the form of production implemented within the firm combined with the inherent educational and skill levels of the work-force which decides whether or not functional flexibility is adopted. Indeed Gertler (1988:424) similarly concludes that, 'outcomes on the shop floor are as much a product of the social/managerial relations surrounding the 'hard' technologies as they are the outcome of the machines themselves'.

The extent to which social/managerial relations within the Massachusetts matched plants were 'predicated on collaboration' (1984:278) were limited. Despite the greater formation of work teams within the Massachusetts firms it was, predominantly, the managers within the Massachusetts matched plants which determined the organization of production and subsequently the degree to which functional flexibility was visible within the firms. While the Massachusetts plants demonstrated that managers determined the manifestation of functional flexibility through the organization of production, within the Northern Ireland plants, lack of co-operation by employees in increasing their skills constrained managers in reorganizing the work flow of production. Therefore, while the results from the matched plants support Piore and Sabel's belief that there must be collaboration between managers and workers in the introduction of flexible production, collaboration did not eradicate managerial control[5].

While Piore and Sabel (1984) emphasise the importance of multi-skilling and increasing functional flexibility, they pay less attention to the emergence of a numerically flexible peripheral work force. Other Flexible Specializationists have however drawn attention to the emergence of less secure, lower paid and lower skilled workers (Storper and Christopherson, 1988). The flexible firm model in particular (Atkinson, 1984) perceives numerical flexibility as being complementary to functional flexibility. Results from the matched plant study along with those of other researchers (Pollert, 1988; Pinch, Mason and Witt, 1991) have demonstrated that there is little evidence supporting this hypothesis. Yet it should be highlighted that within the more labour intensive industry of food processing a substantially higher proportion of numerically flexible workers was found. Yet, this higher incidence within food processing did not bear any relation to the technology within this industrial sector but rather was a reflection of the inherent characteristics of low-skilled, routine and repetitive tasks. Within Fordist production, firms adopting mass production principles also displayed similar job characteristics, yet numerical flexibility was not found within these cases. Therefore the conclusion must be made that the active utilization of a policy of numerical flexibility within firms is not directly related to the technology but instead reflects broader economic factors such as unemployment, social conditions such as the willingness of women to participate within the work force and political conditions such as the availability of welfare to those people not wishing, or unable to actively participate within the economy.

It can be concluded, therefore, that there was little support for the various elements of flexible specialization within the matched plant analysis of small engineering and food processing firms in Northern Ireland and Massachusetts. While this strategy of competition may be manifest within regions such as the Third Italy, it is less applicable to the Northern Ireland or Massachusetts economies. Within these regions, economies of scale are vitally important and the industrial sector is still strongly influenced by the remnants of the mass production industries. Subsequently, small firms are unable to make large capital investments of flexible technology and this accentuates the view that economies of scale are still essential to the existence of the small firm sector. Where economies of scope may provide small firms with a competitive advantage over larger firms, in the absence of economies of scale, small firms will be unable to be cost effective or compete in the same markets as large firms. The technological determinism of the Flexible Specialization theory fails to account for organizational and performance disparities between the matched plants and therefore acts as a poor analytical model against which changes in the flexibility of labour and production can be understood.

The evidence of flexible accumulation within the matched plants

To the Regulationists, flexible accumulation 'is marked by a direct confrontation with the rigidities of Fordism. It rests on a startling flexibility with respect to labour processes, labour markets, products and patterns of consumption. It is characterized by ... greatly intensified rates of commercial, technological and organizational innovation' (Harvey, 1988:8). Within the production process, flexible accumulation would be evident in the introduction of both process and product innovations. Like Flexible Specialization, new technologies in the form of microprocessor technology and microprocessor-based innovations ranging from numerically controlled tools to computer aided design (CAD), computer aided manufacturing (CAM) and robotics would be vital in overcoming the limitations to productivity of Fordist production (Palloix, 1976). Economies of scope would also be important through the attainment of small batch production. Buffer stocks would be virtually eliminated due to the new technology with quality control becoming an integral part of the production process, as opposed to an end-function. Defective parts would therefore be identified at an early stage where reworking was still possible and the overall reject rate of production would be reduced. The gaps within the working day would also be reduced within a flexible system, with the formation of semi-autonomous work teams and job-sharing (Aglietta, 1979; Blackburn et al., 1988), ensuring the maximum use of flexible labour with flexible capital. Finally, with respect to production, (quasi) vertical integration or disintegration would increase in importance with firms benefiting from the formation of strategic alliances.

The results from the matched plant analysis of small engineering and food processing firms in Northern Ireland and Massachusetts have demonstrated that evidence for the component parts of flexible accumulation has varied between the regions. However some similarity in the incidence of flexible accumulation was apparent amongst the food processing matched plants. Within the small food processing firms small batch production was not evident, nor was the attainment of economies of scope with economies of scale dominating the production decisions of the firms. While quality control within this industrial sector was mainly performed as a separate function there was slightly greater evidence within the Massachusetts firms of workers eliminating defective products during the production process as opposed to assuming that supervisors would undertake this role.

Despite production within the food processing firms being dominated by short-runs and high turnover rates of production which lead to low levels of buffer stocks, this was more a reflection of sector-specific market demands than a movement from Fordist to flexible principles of production However evidence with respect to quality control would suggest that the Massachusetts firms are embracing the concepts of flexibility to a greater extent than the Northern Ireland matched plants.

Similarly, the Massachusetts engineering firms displayed greater evidence of flexible accumulation than the Northern Ireland engineering firms. Unlike the food processing firms, where advances towards a more flexible production system were manifest within the engineering sector these were not sector specific characteristics but instead represented attempts to attain a more flexible production process. The Massachusetts firms encouraged the adoption of Just-In-Time (JIT) forms of production along with the implementation of Total

Quality Management (TQM) and Quality Control Circles (QCCs). This led to a widespread elimination of lines of demarcations within the firms with the development of both work teams and job-sharing. Workers within the Massachusetts firms therefore monitored each other and assisted in reducing the porosity of the working day. Alternatively within the Northern Ireland firms production was more characteristic of Just-In-Case production than that of Just-In-Time. Clear job demarcations were evident throughout the shop floor with a clear hierarchy of managerial control existing. Like the food processing firms, quality control within the Northern Ireland firms was mainly performed as a separate function with supervisors assuming this function.

The results also demonstrated that the formation of quasi-vertical integration was more apparent within the Massachusetts firms. A number of the engineering firms established strategic alliances with larger firms and therefore benefited from this in the form of assistance with design, R&D, marketing and technological developments. Additionally, due to the close proximity of these relationships and the concentration of engineering firms within Massachusetts (most notably along Route 128, outside Boston), agglomerative economies were realized by the manufacturing firms. Conversely, within Northern Ireland although it was found that a number of firms undertook sub-contracting with larger firms, the majority of these firms were based in either England, France or Germany. Therefore, the potential to establish strategic alliances were constrained, and thus the traditional relationship of buyer and supplier prevailed. In addition, because, of the geographical distance between these firms it was not possible to benefit from agglomerative economies. Therefore, the Northern Ireland firms were disadvantaged with respect to their ability to benefit from strategic alliances and the associated agglomerative economies.

While it could be stated that the Northern Ireland firms were reluctant to adopt flexible forms of production, it would be more accurate to postulate that the extent to which the firms within Northern Ireland and Massachusetts implemented flexible accumulation was a direct function of the capability and flexibility of the workers. For example, the matched plant analysis demonstrated that the Massachusetts employees had higher levels of educational qualifications and skills than their Northern Ireland counterparts. Additionally this was compounded by the higher investment within Massachusetts firms in the training of their workers, which further increased the potential for functional flexibility to be introduced. Therefore there was a direct and positive relationship between flexibility within the production process and labour flexibility.

While Harvey (1989) accentuates the development of a core and peripheral work force within the flexible accumulation system, the evidence for such within this study was limited. Within the Massachusetts firms, where flexibility was more apparent both within the production process and the labour force, it would have been anticipated that a core and peripheral work force was present.

The results however, illustrated that the evidence for this was limited. For example, numerical flexibility was not found to be significant within any of the firms, yet where it was found it applied to low-skilled workers.

With regard to the educational and skill levels of the employees there was a lower percentage of the total workers within Massachusetts who had no formal qualifications as compared to the Northern Ireland sample. As the Massachusetts engineering firms had adopted flexibility within the production process to a greater extent than the Northern Ireland firms, it would have been anticipated that the managerial and professional workers would have been more qualified than the Northern Ireland workers, with the semi-skilled and un-skilled workers being less qualified than their Northern Ireland counterparts. This was not found and instead, for each of the skill groups the Massachusetts employees were both better qualified and received a significantly greater amount of training than the Northern Ireland employees. However, within the Massachusetts engineering firms the amount spent on training varied between occupational groupings, with managers and technical staff having greater amounts spent on their training than shop floor workers. Although some evidence was found to support the concept of a core and peripheral work force the presence of this was not as extensive as that outlined by Harvey (1989) and it did not lead to a polarisation of workers within the firms.

Finally, with respect to the evidence for flexible accumulation within the matched plants, it can be concluded that as a reflection of the unique conditions of the regions studied, the flexible accumulation system has been adopted at a differential rate. Indeed this is in line with ideas from Regulation theorists who propose that, 'the specific shape of capitalist accumulation can vary considerably over both time and space' (Lipietz, 1987 as quoted in Peck and Tickell, 1992:6). The present period is perceived as being 'embryonic' in the development of a flexible regime of accumulation and therefore while flexibility is seen as the emerging system from that of Fordism and the Keynesian state, theorists have been hesitant to attribute specific characteristics to it. The results from the Northern Ireland, Massachusetts matched plant analysis have reinforced this lack of definition of the neo-Fordist/post-Fordist era. By applying the matched plant methodology in the analysis of flexible accumulation, both regional and sectoral disparities have been identified in the nature of accumulation. In addition, competitiveness gaps and skills gaps have been identified as major constraints in the performance of the small firm sector of the Northern Ireland economy.

While the regulationist theory permits incorporation of the state in the overall framework of economic performance and development, this is not facilitated within the flexible specialization thesis. Significantly, different approaches have been adopted by the states in the promotion of small firms within Northern Ireland and Massachusetts. Within Northern Ireland, while a more proactive role has been adopted by the state in many cases this has veiled the failures of the local economy, with financial assistance concealing inefficiencies and low levels of profitability (see Hitchens, Birnie and Wagner, 1989). Additionally, the inability of Northern Ireland firms to benefit from agglomerative economies severely restricts their potential to adopt flexibility within the structures of the firms.

Within Massachusetts the role of the state within the small firm sector has evolved in collaboration with the universities, unions and businesses of the region. Within Massachusetts a selective programme of targeting assistance at those industrial sectors demonstrating the greatest growth potential, has been adopted. From this the state government in conjunction with other interested parties has formulated support organizations with functions ranging from exporting, marketing, design, R&D, venture capital and education and training. Therefore, within Massachusetts the state has assumed a more flexible format in the area within which it operates, as compared to the traditional Keynesian form of assistance as provided by the Northern Ireland agencies. Yet, with the change in emphasis of the Northern Ireland agencies away from direct financial assistance towards less tangible forms of small firm support, this will encourage firms to reassess their competitive position and strategically target resources towards those areas of their firms which are under-performing. It will however, take a considerable period of time before the implications of this policy can be fully evaluated and the success of this change in emphasis accurately analysed.

It can be summarized, therefore, that while there is evidence of flexibility within the firms, with this being more apparent within Massachusetts as opposed to Northern Ireland, it is certainly not characterized by flexible specialization. Additionally, to state that the evidence of flexibility was incorporated within the flexible regime of accumulation as proposed by the Regulationists would also be incorrect. Thus in the words of Hudson (1989:5) it can be concluded: 'maybe flexibility for some but not flexible accumulation'.

There are clear indications that flexibility has evolved within some of the industrial sectors, namely engineering within the Massachusetts region, and that the adoption of more efficient forms of production has been introduced in an attempt to maintain competitiveness. Yet, if this flexibility is assessed in terms of the criteria as outlined by the Regulationists (Moulaert and Swyngedouw, 1989:336) then clearly it falls short of the true manifestation of flexible accumulation. Therefore, for these firms it is flexibility, but not flexible accumulation.

Evaluation

The matched plant research was designed to establish the extent to which the rhetoric of corporate flexibility was manifest in small manufacturing firms in Northern Ireland and Massachusetts. The results from the matched plants imply that while flexibility was apparent it was not being universally applied, with both regional and sectoral differences being found. Yet, the results suggest that those firms which had adopted labour and production flexibility to a greater extent had also experienced higher turnover growth over the 1988 to 1993 period. The Massachusetts firms within both the engineering and food processing sectors, not only had higher levels of turnover per employee in 1988, but they also demonstrated significantly higher levels of growth between 1988 and 1993.

Measures of growth by changes in employment between 1988 and 1993 did not however reflect changes in turnover growth. For example, within the Northern Ireland firms turnover growth was translated into disproportionate increases in employment, relative to the Massachusetts firms. These results were due to three factors. Firstly output per employee was higher within the Massachusetts firms and had increased by a larger margin over the 1988 to 1993 as compared to the Northern Ireland firms. Secondly, lower wage rates in Northern Ireland, relative to Great Britain over the 1988 to 1993 period encouraged investment in labour intensive industries as opposed to capital intensive industries (Gudgin, 1994). Thirdly, government agencies within Northern Ireland promoted job creation in response to company assistance and thus created 'a vertical maternalistic relationship between the enterprise and government agencies' (Clulow and Teague, 1993:105).

While there appeared to be greater cultural awareness within Massachusetts of the need to organize production flexibly, Northern Ireland companies were less conscious of changing market demands. It is argued that this is due to the peripherality of Northern Ireland relative to the European core. Peripherality negatively affecting the performance of firms through higher costs of transportation[6], fewer opportunities to form strategic alliances, a constrained external labour market and a risk adverse environment (Clulow and Teague, 1993:109). Examination of the regional economies suggest that there is a positive correlation between the organizational aspects of the firm and the environment within which the firms operate. The Massachusetts matched plants, in particular, illustrated a distinctive link between the characteristics of the firms and their regional support environment. Although the Massachusetts economy has been transformed since the mid 1970s with the evolution of the high-technology sector, it would be inaccurate to suggest that all developments within the manufacturing sector were attributable to the positive externalities of the high technology sector. Instead, a large proportion of the manufacturing sector has remodelled the governing principles of its productive system in line with the demands of the new competitive environment (Carnevale, 1991).

While it is impossible to assume representativeness of the small firm sample in Northern Ireland to the population of small firms as a whole, a number of differences can be identified and the implications of these in terms of the measures which the sample of Northern Ireland firms could take, outlined.

The evidence of numerical flexibility within the matched plants was somewhat limited. Neither the Northern Ireland or Massachusetts firms displayed significant levels of part-time or temporary workers. However, there was a slightly greater propensity for the food processing firms to have part-time or temporary workers than the engineering firms. While workers may have specific skills on joining a firm, it is only through employee training by the firm that job-specific skills can be obtained. As firms are increasingly forced to invest in the training of employees - even if this is only in the form of work-shadowing and on-the-job training of employees - there will be a reluctance to release employees during less prosperous periods of production due to the costs associated with (re)training new employees when demand increases. Consequently, it can be concluded that there is a direct relationship between the ability of firms to adopt a policy of numerical flexibility and the skill levels of the work force. Therefore while some sectors may benefit from a ready supply of labour in the external labour market this will be largely confined to the low-skilled sectors.

Greater evidence for functional flexibility was found within the Massachusetts matched plants. The ability to deploy workers flexibly throughout the workplace was pre-determined by the competence levels of the employees. In general the Massachusetts employees had higher qualification levels and received more training than their Northern Ireland counterparts. In addition, the nature of employee training differed between the regions with the Massachusetts employees being encouraged to acquire both cognitive and job specific skills. In contrast, while Northern Ireland employees received substantially lower levels of investment in their training, this was confined to job-specific functions. The skill enhancing of Massachusetts employees facilitated team working and the formation of quality control circles which had positive effects upon the efficiency and the reject rate of production.

While the Massachusetts firms sought to improve the skill levels of all of their employees, where training was undertaken within the Northern Ireland firms this was largely confined to managers. The Northern Ireland firms could therefore benefit from increasing the skill and competence levels of *all* employees as opposed to just a few. Further, the introduction of regular meetings/progress briefings within work groups/teams in the Northern Ireland firms might encourage a sense of community amongst the workers. As was found within the Massachusetts firms, this would also improve employee awareness of cross-functional activities.

The enhancing of employee skills within Massachusetts led to a reduction in the lines of job demarcation. The results suggest that if the Northern Ireland firms were to encourage interdependent relationships amongst workers, this would lead to improvements in efficiency as wastage would be reduced through the formation of quality control circles. Further, in improving productive efficiency supervisors could be encouraged to develop skills which increase the innovative capacity of the firm and reduce the extent to which they act as trouble-shooters and overseers. Indeed, despite the similarity between the Northern Ireland and Massachusetts firms in the age and technological sophistication of capital, the skill levels of the Massachusetts employees were facilitating the capital to be used much more efficiently.

While it was found that the Northern Ireland small firms were characterized by greater informality and lack of daily planning, this factor could be eradicated with greater attempts being made by managers to construct formal work plans and specified deadlines. Further, Northern Ireland firms could reduce their dependency upon the local market, either through the establishment of external contracting relationships or the penetration of new markets. It is believed that both of these measures would make a positive contribution to the competitiveness of Northern Ireland firms.

In terms of flexibility within the production process, the matched plant findings suggest that the Massachusetts firms had adopted Just-In-Time principles of production in an attempt to reduce costs and respond rapidly to changing market demands. Many of these small firms had also captured niche markets either through their products or services, in an attempt to increase market share in an increasingly competitive environment. Where this was not possible, 'offensive enterprise calculations' (Clulow and Teague, 1993:93) had been embraced through the development of innovative capability and R&D (or where this has not been economically viable within the firm, strategic alliances were formed which enabled firms to undertake R&D through co-operation with larger enterprises were formed).

The high investment in the skill levels of employees to encourage functional flexibility throughout the firm and the adoption of technologically advanced capital to enable the introduction of flexible production had also been widely adopted within the Massachusetts firms. Yet, the research demonstrated that irrespective of the changes in employee skills, a firm will not increase its competitive performance in the absence of positive managerial attitudes towards strategic and organizational change. The aggressiveness and assertiveness of the Massachusetts managers as compared to the complacency of the Northern Ireland managers is one factor which was apparent from the interviews. While the reasons for this are difficult to define accurately and quantitatively, entrepreneurial culture has been allied with the political instability within Northern Ireland along with high levels of government intervention through subsidies and grants.

An environment of 'competition and co-operation' could be nurtured within Northern Ireland akin to that in Massachusetts. Within Massachusetts 'competition and co-operation' have been encouraged through the development of 'Centres of Excellence' which 'consciously support the interstate and international competitiveness of strategically important sectors' (Ferguson and Ladd, 1988:77). The Northern Ireland development agencies could benefit from targeting its assistance more effectively towards 'strategically important sectors' (ibid). While steps have been made by the Northern Ireland agencies to reduce the level of direct financial assistance to firms and instead encourage the development of less tangible functions within the firms, this has fallen short of providing the catalysts whereby an economic environment of competition and co-operation can be encouraged. Yet, it should be acknowledged that Northern Ireland does not have an industrial base capable of self-sustaining growth to the same extent as found within Massachusetts, for example, in polymers, biotechnology, marine sciences and photovoltaics (solar energy). Yet, if the Northern Ireland agencies placed greater emphasis upon the specialist talent within the universities through the development of Centres of Excellence then it is possible that industrial and economic spin-offs could be attained which could be translated into viable commercial activities within the province, through the continued support of the agencies.

This research has not sought to propose that the Northern Ireland firms adapt their methods of production to imitate those of the Massachusetts firms. Indeed, the uniqueness of the two regions in terms of their historical development combined with their economic, spatial, social and political characteristics have cultivated diverse landscapes with individual determinants of change. However, this does not imply that Northern Ireland firms should ignore the source of their competitive weaknesses. While the uniqueness of the Northern Ireland economy has to be accepted, any analysis of the regions should be placed within a wider global context of which it is an integral part. Development of the local economy must act as the crux upon which policy is formulated, with growth being perceived as an attainable and beneficial consequence, but not as an end in itself. While regional economic growth in terms of increased output is desirable, regional development can only be achieved through changes in the institutional structures and organizational frameworks which constitute the regional economic milieu.

Notes

1 That is, while the majority of shop floor workers within clothing firms in Germany could efficiently perform five or more functions, only 5 to 10 per cent of the Northern Ireland shop floor workers could perform this number of functions.

2 It should be noted that the Regulationists identified the structural problems of mass production as being growing unrest amongst labour and technical problems within production.

3 The age of capital equipment is significant in that newer equipment is more efficient and therefore positively affects productivity. However, this is offset by the higher level of depreciation which adversely affects profitability, and the higher value of the fixed assets which adversely affects the return on assets ratio. On balance the lower the age of the capital equipment the better the performance of firms (McFerran, Hewitt-Dundas and Roper, 1996).

4 It should be noted that even within the Regulationist's perspective disagreement exists concerning the successor to Fordism. For example while the 'Neo Fordists' such as Aglietta (1979), believe that new technology will lead to the deskilling of workers, Post-Fordists, such as Kaplinsky (1984) have highlighted that skills will be upgraded in line with changing technological demands (see Wood, 1989).

5 The maintenance of managerial control is cited as being central to the continuation and success of neo-Fordism in the short term (Aglietta, 1979).

6 Contrary to this perception of geographical peripherality being associated with higher transportation costs, Hitchens It should be noted that even within the Regulationist's perspective disagreement exists concerning the successor to Fordism. For example while the 'Neo Fordists' such as Aglietta (1979), believe that new technology will lead to the deskilling of workers, Post-Fordists, such as Kaplinsky (1984) have highlighted that skills will be upgraded in line with changing technological demands (see Wood, 1989).

6 The maintenance of managerial control is cited as being central to the continuation and success of neo-Fordism in the short term (Aglietta, 1979).

6 Contrary to this perception of geographical peripherality being associated with higher transportation costs, Hitchens and Birnie (1989) have presented evidence which fails to support this hypothesis.

Appendix: Accumulation in the 20th Century

Regimes of accumulation have been described as either intensive or extensive, being determined by the scale of production, the organisation of work and the substitution of labour to capital. Aglietta (1979:71) identifies an extensive and an intensive form of accumulation. Within the extensive regime of accumulation, relative surplus value is obtained through the transformation of the organisation of labour, which is supported through mechanisation. Accumulation occurs through the building-up of layers of industry. Recurrent difficulties are however encountered throughout the process which, depending on their severity, may generate a reorganisation of the regime of accumulation, either through a new accumulation system, or a new mode of social regulation, or, indeed, the renewal of both. The intensive regime of accumulation differs from the extensive regime, in that the former creates a new mode of life for those possessing labour-power through the establishment of a logic which operates on the totality of time and space occupied or traversed by its individuals in daily life. A consumption norm is therefore established based upon utilitarianism, and is stratified within society in conjunction with the stratification of the social groups. Thus a rapid increase in surplus-value is attainable through the integration of the different economic groups.

Attempts have been made to periodise the economic history of the 20th Century in terms of regimes of accumulation. Through an examination of the processes at work within society a number of intervals have been identified as having a specific form of accumulation system, an associated mode of social regulation, and a period of crisis through which a new regime has been initiated. The periodisation of accumulation in the 20th Century is illustrated in Table A1.1.

Table A1.1
The regimes of accumulation throughout the 20th Century

Period	To 1914	1918-1939	1945-1973	1974- Present
Accumulation System	Extensive	Emerging Intensive	Intensive (Fordist)	Emerging Flexible? Protracted Crisis?
Mode of Social Regulation	Competitive	Crisis of Competitive	Monopolistic (Fordist-Keynesian)	Crisis of monopolistic, Neo-Competitive? Neo-Corporatist? Neo-Conservative?

Source: Peck and Tickell (1992:9)

The first period identified extends from the end of the second industrial revolution c.1870[1] to approximately 1914, or the commencement of the First World War. Throughout this period, the capitalist system sought to narrow the gap between relative and absolute surplus-value. This it achieved most successfully through the combination of an extensive accumulation system and a competitive mode of social regulation. The key to this was the widespread introduction of mechanisation into the workplace with workers becoming appendages of the machines. Tasks which were previously performed by workers and craftsmen, were now being carried out through the mere application of human labour to the technology. Carnevale (1991:5) states that the "natural rhythms characteristic of the craft and farm economy bowed to bureaucratic procedure and the machine cadences of the factory." Technology provided the basis around which labour was organized in the most efficient manner.

Care should be taken to avoid over emphasising the power of technology with respect to the organisation of labour, and any form of technological determinism must be rejected. Indeed, Aglietta (1979) points out that the technical division of labour is not absolute and abstract from other forces, but instead is heavily impinged upon by the social division of labour. Therefore, both the technical and social divisions of labour are interdependent aspects, and must be interpreted as such. Nevertheless, it must be acknowledged that the system of mechanisation during the period of extensive accumulation reduced labour to a position where its output norm was determined in terms of time, a feature which led to the collectivization of labour-power as an homogeneous variable generating the surplus-value necessary for the reproduction of the circuit of capital.

This system of accumulation is most commonly referred to as Taylorism, having been pioneered by F.W Taylor, through "The Principles of Scientific Management" (1911). The primary objective of Taylorism was to increase the level of surplus-value to that approaching an absolute state. As aforementioned, through the exploitation of labour, namely, the concentration and increasing intensity of work, along with continual supervision, labour productivity was increased to a level approaching maximum efficiency. For example, the introduction of conveyor belts reduced time wastage as parts were brought to the workers throughout the productive cycle. Additionally, conveyor belts dictated the pace of work, and their continuity or semi-continuity dramatically reduced rest-time for the workers. Hence Taylor proposed that the emerging technology of the late 19th Century could be easily incorporated into the workplace through the redesigning of jobs. Therefore, the productive cycle was broken down into a number of stages, which were then further fragmented into tasks which could be fulfilled by the workers.

Although Taylorism was first introduced to the engineering industry within the United States at the end of the 19th Century, Taylor's ideology was rooted in the much earlier works of Ure and Babbage in the mid-19th Century and Gilbraith in the 1890s. Aglietta (1979:115) states that it was a "capitalist response to the class struggle in production in a phase when the labour process was composed of several segments, each organized on mechanical lines internally, yet whose integration still depended on direct functioning of the output norms."

With respect to the competitive mode of social regulation, this was manifest through a non-interventionist policy by the nation-state, whereby individual firms or industries, negotiated wage rates under a minimum level of government control. Subsequently, variations and uncertainty in the demand for commodities were translated to the workers' wages, with them being unable to intervene. Yet for the capitalist this provided stability in the level of direct real wages which were paid (Lipietz 1986:25). Despite the advantages for the capitalists, it was these factors which led to the capitalist crisis at the end of the Second World War.

During the period of extensive accumulation it was primarily the leading industrial sectors producing capital goods which were the major beneficiaries of the regime. However, following the First World War, Taylor's principles were translated to consumer goods industries as capitalists sought to expand their share of the market and their levels of accumulation. Thus, although productivity gains rose dramatically, these were extraneous without a complementary rise in the levels of demand. Within the prevailing competitive mode of social regulation wage levels were unable to respond to increased levels of productivity, and thus, the boom generated through the intensive extraction of surplus-value was obsolete as it induced a crisis of overaccumulation.

A fresh restructuring and coupling of the accumulation system and the mode of social regulation was necessary, and hence a monopolistic mode of social regulation emerged following the end of the Second World War. With unprecedented productivity gains in both capital and consumer goods industries, due to technical changes and the deskilling of the workforce, mass production was soon combined with mass consumption, as the higher wages paid for monotonous repetitive tasks facilitated the widespread purchase of standardized goods.

This accumulation system was known as *Fordist* with the respective mode of social regulation being known as the *Fordist-Keynesian mode*. Within the United States, the foundation of the new mode of social regulation was attributed to President Roosevelt's *New Deal for the American People* along with the rising power of the trade unions (Aglietta 1979). Unions became more significant as the exploitation of labour was intensified, with the refinement of Taylor's principles to approach absolute surplus-value.

Harvey (1989:125) argues that the major factor differentiating Fordism from Taylorism was the vision and the "explicit recognition that mass production meant mass consumption, a new system of the reproduction of labour power, a new politics of labour control and management, a new aesthetics and psychology, in short a new kind of rationalized, modernist, and populist democratic society." Ford therefore believed that corporate power within society was sufficient to revolutionise the type of society present and the people therein.

Despite the successful adoption of Fordism, it was not without its problems and the perception of a universal regime of accumulation under the principles of Fordism must be revoked. For example, outside America, labour organisations were often too strong to accept this new theory of labour and further, immigration, which played a major role in the acceptance of the Fordist work practices in the United States, was of limited benefit within Europe[2].

Particularly within Europe, the technical division of labour was not characterized by a horizontal flow of the production process but rather, was preoccupied with organisational structures and a hierarchical ordering of authority, responsibility and information flows, as influenced by the work of Henri Fayol's "*Administration industrielle et générale*" (1916). Harvey (1989) states that it was not until the 1950s that Fordism was fully established in Europe. Nevertheless, succeeding its implementation, it facilitated a long post-war boom until its inherent contradictions led to a crisis period in the 1970s. Throughout these years economic rates of growth in the advanced industrialized countries were relatively stable and crisis tendencies were kept to a minimum. Within this intensive regime of accumulation, the accumulation system was characterized by an increase in productivity, along with a higher level of fixed capital per capita. It is evident that productivity values reached levels which were never before possible, and indeed since 1974 have not been achieved in the Western economies. The mode of social regulation constituted a situation of continual adaptation to increased productivity gains. The Keynesian practice of governments, displayed through the formation of a welfare state in the UK, enabled the continuation of high levels of consumption, irrespective of the employment status of individuals. Further, trade unions began to determine wage levels on a collective basis, an element which reflected the rising power of the workers at that time. Indeed, the power to initiate wage rises was a contributory factor in the decline of Fordism in the 1970s.

The beginning of the crisis within the intensive regime of accumulation is assigned to the late 1960s (Peck and Tickell 1992:9) when productivity in the propulsive industries began to slow as the technical limits were met, yet wage levels persisted to rise due to increased collective bargaining power of the trade unions. Further, levels of investment and capital intensification reduced and major problems arose with respect to labour relations, as demonstrated by the strikes and labour disruptions between 1968 and 1972 in the United States.

The rate of profit declined and industries became laden-down with debt, of which the only suitable amelioration was to internationalise production and shed expensive labour at home. For the governments of these countries this led to an increase in the percentage of the population dependent upon the welfare state. Aglietta (1979:122) concludes that; "the class struggles in production today bear the germ of a major new transformation of the labour process - Neo-Fordism", a regime which is engrossed by further intensification and greater socialisation of collective consumption. At such suggestions the aims of regulation theory must be re-echoed in that it does not seek to solve crises which emerge but rather, to "provide a scientific interpretation of the appearance of novelty in the social realm." (Aglietta 1979:37).

The response of government, most notably within the US and the UK to these escalating demands upon their resources was to adopt monetary policy. Yet this brought with it problems in the form of inflationary pressures which eventually brought the crisis of Fordism to a climax. The neo-Conservatist attitudes which emerged at that time as encapsulated within the ideologies of the Thatcher and Reagan administrations, were fundamental in channelling the crisis of the Fordist regime. Previous to this the Nixon and Heath administrations had encountered the mounting pressures of the Keynesian-Fordist regime of accumulation, and this had led to rising struggles with organized labour and the retrenchment of state expenditures. Thus, the progressive withdrawal of support for the welfare state by the Conservative government, and their attack upon real wages and the power of the unions, was manipulated by the Conservative administrations to their advantage, being viewed as "administering powerful doses of unpalatable medicine to restore the health of ailing economies." (Harvey 1989:168).

The question therefore arises, as to what form the economic restructuring to Fordism took, or as we are presently living within the post-Fordist regime, what form is it ascribed today? Harvey (1989:142) highlights the predominant feature of Fordism, as being bound by "rigidity…rigidity of long term and large scale fixed capital investments in mass production systems that precluded much flexibility of design and presumed stable growth in invariant consumer markets".

Any attempts prior to the crisis of the early 1970s, to overcome these rigidities encountered class struggles and revolt from the working classes. Yet, within the new regime of accumulation 'flexibility' has become the buzz-word, found throughout the literature not only from proponents of a flexible regime of accumulation such as Harvey (1989); Peck and Tickell (1991,1992); Aglietta (1979); Moulaert and Swyngedouw (1989) etc., but also amongst neo-corporatists such as Lash and Bagguley (1988); and others such as Gertler (1987); Morris (1988) and Hudson (1989). Although caution is demanded by some in determining that a transition has occurred to a flexible regime due to the need for a transitory period, like that found from c.1914 to 1945, with respect to the movement from an extensive to an intensive regime of accumulation, over 20 years has now passed since the start of the present crisis, and therefore it would be anticipated that the period of transition is approaching completion. The main characteristics of the flexible regime of accumulation are examined more thoroughly in Chapter One.

Notes

1 It should be noted that in a number of trades the introduction of mechanisation as the controlling force within the production process occurred prior to 1870, however, this date is interpreted as being representative of the introduction of such technology within industry as a whole.

2 For an examination of why Fordism was never fully implemented in the United Kingdom, see Nolan and O'Donnell, 1991.

Bibliography

Abraham, K. (1987), 'Flexible staffing arrangements and employers' short-term adjustment strategies', in Hart, R.(ed.), *Employment, unemployment and labor utilization,* Unwin Hyman: Boston.

Acs, Z.J. and Audretsch, D.B. (1991), *Innovation and technological change: an international comparison*, University of Michigan Press: Ann Arbor.

Adams, A. (1982), 'Barriers to product innovation in small firms: Policy Implications', *European Small Business Journal,* Vol.1, No.1, pp. 67-86.

Aglietta, M. (1979), *A theory of capitalist regulation: The US experience*, New Left Books: London.

Allen, J. and Massey, D. (1988), *The economy in question*, Sage Publications, in association with the Open University Press: London.

Altshuler, A., Anderson, M., Jones, D., Roos, D. and Womack, J. (1984), *The future of the automobile*, MIT Press and Allen and Unwin: Cambridge, Massachusetts and London.

Amin, A. (ed.), (1994), *Post-Fordism - a reader*, Blackwell: Oxford, Cambridge Massachusetts.

Armingtom, C. and Odle, M. (1982), 'Small business - how many jobs?' *Brookings Review*, Winter, pp.14-17.

Arrighi, G. (1990), 'The development illusion: A reconceptualization of the semiperiphery', in Martin, W.G. (ed.), *Semiperipheral states in the world economy*, Greenwood Press: New York, pp.11-42.

Atkinson, J. (1984), 'Flexibility, uncertainty and manpower management', *Institute of Manpower Studies*, Report 89, Falmer: University of Sussex.

Atkinson, J. (1985), 'Flexibility: planning for an uncertain future', *Manpower Policy and Practice*, Vol. 1, Summer, pp. 26-29.

Atkinson, J. (1988), 'Recent changes in the internal labour market structure in the UK' in Buitelaar, W. (ed.), *Technology and work: Labour studies in England, Germany and the Netherlands*, Avebury: Aldershot, pp. 133-149.

Aydalot, P. (1986), 'Trajectories technologiques et milieux innovateurs', in Aydalot, P. (ed.), *Milieux innovateurs en Europe*, GREMI, Paris, pp. 345-361.

Aydalot, P. and Keeble, D. (1988), 'High technology industry and innovative environments in Europe: an overview', in Aydalot, P.H. and Keeble, D. (eds.), *High technology industry and innovative environments: the European experience*, Routledge: London, pp.1-21.

Bannock, G. and Daly, M. (ed.) (1994), *Small Business Statistics*, Paul Chapman: London.

Barkham, R. (1992), 'Entrepreneurial characteristics and the size of the new firm: A model and econometric test', Paper presented at the International Conference on Birth and Start up of Small firms, Boconni University: Milan, June 11-19.

Barkham, R., Hart, M., Hanvey, E. and Gudgin, G. (1994), *Determinants Of small firm growth: An interregional analysis for the UK*, Jessica Kingsley: London.

Baumol, W.J. (1989), 'Entrepreneurship - Productive, unproductive and destructive', *Journal of Political Economy*, Vol. 98, No. 5, pp. 893-921.

Beasley, M.E. and Wilson, P.E. (1981), *Government aid to small firms in Britain*, UKSBMTA Conference Paper, London.

Beaumont, J.A. (1982), 'Aspects of a changing relationship between food manufacturing and distributors in the UK', Paper presented to OECD Conference, The adjustment and the challenge facing the food industries in the 1980s, Paris.

Belous, R.S. (1989), *The contingent economy: The growth of the temporary, part-time and subcontracted workforce*, National Planning Association, Washington D.C.

Bennett, R.J. and McCoshan, A. (1993), *Enterprise and human resource development - local capacity building*, Paul Chapman: London.

Best, M.H. (1990), *The New Competition: Institutions of Industrial Restructuring*, Polity Press: Cambridge.

Binks, M. and Coyne, J. (1983), *The birth of enterprise: An analytical and empirical study of the growth of small firms*, Hobart Paper 98, Institute of Economic Affairs: London.

Birch, D.L. (1979), *The job generation process*, MIT Program on Urban and Regional Change, Cambridge: Massachusetts.

Blackburn, P., Coombs, R. and Green, K (1985), *Technology, economic growth and the labour process*, Macmillan: London.

Blackburn, R.A. (1990), Book review, Amin, A. 'Flexible Specialisation and small firms in Italy: myths and realities', *International Small Business Journal*, Vol. 8, No. 2, pp. 58-60.

Bluestone, B. and Harrison, B. (1982), *The de-industrialization of America*, Basic Books: New York.

Bolton Report (1971), Interim report by the committee to review the functioning of small firms within the economy, Cmnd 4811, HMSO: London.

Borooah, V.K. (1993), 'Northern Ireland - Typology of a regional economy', in Teague, P. (ed.), *The economy of Northern Ireland*, Lawrence and Wishart: London, pp.1-24.

Boyer, R. (ed.) (1986), *La Flexibilite du travail en Europe*, Editions La Decouverte: Paris.

Boyer, R. (1988), *The Search for Labour Market Flexibility: The European Economies in Transition*, Clarendon Press: Oxford.

Boyer, R. (1990), *The Regulation School: a critical introduction*, Columbia University Press: New York.

Boyer, R. (1991), 'The politics of flexibility', in Jessop, B., Kastendiek, H., Nielsen, K. and Pedersen, O.K., *The politics of flexibility - Restructuring state and industry in Britain, Germany and Scandanavia*, Edward Elgar: New York.

Boyer, R. and Mistral, J. (1983), *Accumulation, inflation, crises*, Presses Universitaires de France: Paris.

Bradley, J. and Wright, J. (1993), 'The determinants of growth and development in the two economies of Ireland', Paper presented at the conference on Growth and Development in the Two Economies of Ireland, Northern Ireland Economic Research Centre, Belfast.

Bramlette, C.A. (1984), 'Free to change', *Training & Development Journal*, Vol. 38, Part 3, pp. 32-40.

Brusco, S. (1986), 'Small firms and industrial districts: The experience of Italy', in Keeble, D. and Egbert, W. (eds.), *New firms and regional development*, Croom Helm: London, pp. 184-203.

Buckland, C. (1993), 'Policy development for the small business sector', Paper presented at the economic change and development conference, Queen's University of Belfast, Belfast.

Bull, P.J. (1984), 'The impact of government policy on industrial diversification: the Northern Ireland case', in Barr, B.M. and Waterns, N.M. (eds.), *Regional diversification and structural change*, Tantalus Research: Vancouver.

Burns, A. (1934), *Production trends in the United States since 1870*, National Bureau of Economic Research.

Buzzell, R.D. and Gale, B.T. (1987), *The PIMS principles*, Free Press and Collier Macmillan: New York and London.

Camagni, R. (ed.) (1991), *Innovation networks: spatial perspectives*, Belhaven Press: London and New York.

Campbell, J. (1992), 'Self Employment: into the 1990s', *Employment Gazette*, pp. 269-292.

Carnevale, A.P. (1991), *America and the new economy*, Jossey-Bass: San-Francisco.
Carr, C. (1990), *Britain's competitiveness: The management of the vehicle components industry*, Routledge: London.
Carson, C.C. (1993), 'Perspectives on education in America - An annotated briefing', *Journal of Educational Research*, Vol. 86, No. 5, pp. 259.
Cecchini, P. (1988), The cost of non-Europe, European Commission.
CEDEFOP (1990), *Vocational training scenarios for some member states of the European Community: A synthesis report for France, Greece, Italy, Portugal, Spain and the United Kingdom,* European Centre for the Development of Vocational Training: Berlin.
Centre for Labour Market Studies (CLMS) (1991), *One hundred years of population and economic changes: The Massachusetts experience 1890-1990*, CLMS, Northeastern University: Boston, Massachusetts.
Clulow, R. and Teague, P. (1993), 'Governance structure and economic performance', in Teague, P. (ed.), *The economy of Northern Ireland*, Lawrence and Wishart: London, pp. 60-121.
Cohen, R. (1981), 'The new international division of labor, multinational corporations and urban hierarchy', in Dear, M. and Scott, A. (eds.), *Urbanization and urban planning in capitalist society*, Methuen: New York, pp. 287-315.
Coyne, J. and Wright, M. (1986), 'The small firm, government policy and industrial change', in Hall, G. (ed.), *European industrial policy*, Croom Helm: London.
Curran, J., Kitching, J., Abbott, B. and Mills, V. (1993), *Employment and employment relations in the small service sector enterprise - A report*, ESRC, Centre for research on small service sector enterprises, Kingston Business School.
Cusumano, M. (1985), *The Japanese automobile industry*, Cambridge University Press: Cambridge Massachussetts.
Daly, A., Hitchens, D.M.W.N. and Wagner, K. (1985), 'Productivity, machinery and skills in a sample of British and German manufacturing plants', *National Institute Economic Review*, No. 111, pp. 48-62.
Daniel, W. and Millward, N. (1983), *Workplace industrial relations in britain*, Heinemann: London.
Dankbaar, B. (1988), 'New production concepts, management strategies and the quality of work', *Work, Employment and Society*, Vol. 2, No. 1, pp. 25-50.
De Vroey, (1984), 'A regulation approach to the interpretation of the contemporary crisis', *Capital and Class*, No. 23, pp. 45-66.
Deloitte, Haskins & Sells (1989), *Management challenge for the 1990s: The current education training and development debate*, Department of Employment: Sheffield.

Department of Economic Development, (1987), *Northern Ireland competing in the 1990s*, Department of Economic Development (Northern Ireland): Belfast.

Department of Economic Development, (1987), *Pathfinder report - Building a stronger economy, The Pathfinder process*, Department of Economic Development (Northern Ireland): Belfast.

Department of Employment and Training, (1990), *Career choice in a changing economy*, Center for Labor Market Studies, Commonwealth of Massachusetts: Boston.

Department of Employment and Training, (March 1991), *Distribution of establishments and employment by size groups*, Department of Employment and Training (Massachusetts), Economic Research and Analysis Service, Boston.

Department of Employment and Training, (May 1991), *Massachusetts service-delivery area job market reports - The Southern Worcester SDA*, Center for Labor Market Studies, Commonwealth of Massachusetts: Boston.

Dertouzos, M.L., Lester, R.K. and Solow, R.M. (1989), *Made in America*, MIT Press: Cambridge, Massachusetts.

Doeringer, P. and Piore, M.J. (1971), *Internal labour markets and manpower analysis*, Heath: Lexington, Massachusetts.

Dore, R. (1986), *Flexible Rigidities*, Athlone Press: London.

Dosi, G., Freeman, C., Nelson, R., Silverberg, G. and Soete, L, (eds.) (1988), *Technical change and economic theory*, Frances Pinter: London.

Drucker, P.F. (1985), *Innovation and entrepreneurship: practice and principles*, Harper & Row: New York.

Dukakis, M.S. and Kanter, R.M. (1988), *Creating the future - The Massachusetts comeback and its promise for America*, Summit Books: New York.

Dun and Bradstreet Corporation, (6/8/92), *Business failures in first half of 1992 rise 17% by 50, 582 compared with first half of 1991*, News Bulletin, Dun and Bradstreet: New York.

Dunford, M. (1990), 'Theories of regulation', *Environment and Planning D: Society and Space*, Vol. 8, No. 3, pp. 297-321.

Dunkelberg, W.C. and Cooper, A.C. (1982), 'Patterns of small business growth', Academy of Management Proceedings 1982, pp. 409-413.

Enderwick, I.P., Gudgin, G. and Hitchens, D.M.W.N. (1990), 'The role of the firm in manufacturing', in Harris, R.I.D., Jefferson, C.W. and Spencer, J.E. (eds), *The Northern Ireland economy: a comparative study in the economic development of a peripheral region*', Longman: London.

Farnham, D. (1990), 'Trade union policy 1979-89: Restriction or reform?', in Savage, S.P. and Robins, L. (eds.), *Public policy under Thatcher*, Macmillan: London.

Fayol, H. (1916), 'Administration industrielle et generale', *Bulletin de la Societe de l'Industrie minerale*, Paris.
Fenton-O'Creevy, M. Nicholson, N. (1994), *Middle managers: Their contribution to employee inVolvement*, Employment Department Research, Series No.28: London.
Ferguson, R.F and Ladd, H.F. (1988), 'Massachusetts', in Foster, R.S. (ed.), *The new economic role of American states*, Oxford University Press: New York.
Fine, B. (1989), *Marx's capital*, Macmillan: London.
Finegold, D. and Soskice, D. (1988), 'The failure of training in Britain: Analysis and prescription', *Oxford Review of Economic Policy*, Vol. 4, No. 3, pp. 21-53.
Florida, R. and Kenny, M. (1990), 'High technology restructuring in the USA and Japan', *Environment and Planning A*, Vol. 22, pp. 233-252.
Foster, A.F. (1988), 'Retraining adults: An examination of national training policy, its influence on the structure and content of adult job retraining in further education and industry, and consequent impact on the experiences and perceptions of participating adults', unpublished PhD thesis, University of Warwick: Warwick.
Fothergill, S. and Gudgin, G. (1982), *Unequal growth: Urban and regional employment change in the UK*, Heinemann: London.
Frankel, L.D. and Howell, J.M. (1985), *Economic revitalization and job creation in America's oldest industrial regime*, Bank of Boston: Boston, Massachusetts.
Freeman, C. (1987), *Technology policy and economic performance: Lessons from Japan*, Frances Pinter: London.
Gaffikin, F. and Morrisey, M. (1992), *The new unemployed*, Zed Books: Belfast.
Gallagher, C.C. and Miller, P. (1991), 'New fast growing companies create jobs', *Long Range Planning*, Vol. 24, No. 1, pp. 96-101.
Gallagher, C.C. and Stewart, H. (1984), *Jobs and the business life cycle in the UK*, Research Report Number 2, Department of Industrial Management: University of Newcastle Upon Tyne.
Gertler, M.S. (1988), 'Flexibility revisited: Districts, nation-states and the forces of production', *Transactions of the Institute of British Geographers*, Vol.17, pp. 259-278.
Gertler, M.S. (1988), 'The limits to flexibility - Comment on the post-Fordist vision of production and its geography', *Transactions of the Institute of British Geographers*, Vol. 13, pp. 419-433.
Giaoutzi, M., Nijkamp, P. and Storey, D.J. (Eds.) (1988), *Small and medium sized enterprises and regional development*, Routledge: London and New York.

Gibb, A. (1987), 'Enterprise culture - Its meaning and implications for education and training', *Journal of European Industrial Training*, Vol.11 No. 2, pp.1-28.

Gilbert, N., Burrows, R.and Pollert, A. (1992), *Fordism and flexibility: divisions and change*, Macmillan Press: London.

Goodhart, D. (24/5/94), *Third way to the top of the pile: David Goodhart weighs the debate on UK competitiveness in the light of disparate global economic models*, Financial Times: London.

Gough, J. (1992), 'Where's the value in 'Post-Fordism'?', in Gilbert, N., Burrows, R. and Pollert, A. (eds.), *Fordism and flexibility: Divisions and change*, Macmillan Press: London, pp. 31-49.

Gudgin, G. (1984), 'Employment creation by small and medium sized firms in the UK', in Greffe, X. (ed.), *Les P.M.E. Créent-Elles Des Emplois?*, Economica: Paris.

Gudgin, G., Hart, M., Fagg, J., D'Arcy, E. and Keegan, R. (1989), *Job generation in manufacturing industry 1973-86: A comparison of Northern Ireland with the Republic of Ireland and the English Midlands*, NIERC: Belfast.

Gudgin, G., Scott, R., Hanvey, E. and Hart M, (1995), 'The role of small firms in employment growth in the two economies of Ireland: Public policy, growth and employment', in Bradley, J. (ed.), *Irish studies in management,* Oak Tree Press: Dublin.

Guidetti, G. (1995), 'Labour flexibility and segmentation: An analysis of the effects of techno-organisational constraints', Department of Economics University of Aberdeen, Discussion Paper 95-10, Aberdeen.

H.M.S.O. (1986), *Working Together: Education and Training*, Cmnd. 9823, Employment Department/Department of Education and Science, HMSO: London.

Hall, P. (1981), 'The geography of the fifth Kondratief cycle', *New Society*, No. 55, pp. 53-57.

Harris, R.I.D. (1991), *Regional economic policy in Northern Ireland 1945-1988*, Avebury: Aldershot, England.

Harris, R.I.D. (1993), 'Retreat from policy: The rationale and effectiveness of automatic capital grants', in Harrison, R. and Hart, M. (eds.), *Spatial policy in a divided nation*, Jessica Kingsley: London.

Harris, R.I.D., Jefferson, C.W. and Spencer, J.E. (eds.), (1990), *The Northern Ireland economy: Economic development in a peripheral region*, Longman: Harlow.

Harrison, B. (1988), 'The economic development of Massachusetts', in Lampe, D. (ed.), *The Massachusetts miracle - High technology and economic revitalization*, Massachusetts Institute of Technology: Cambridge, Massachusetts, pp. 74-102.

Harrison, B. and Bluestone, B. (1988), *The great U-turn: Corporate restructuring, laissez faire and the rise of inequality in America*, Basic Books: New York.

Harrison, R. (1991), 'Marxism and social-democracy - the revisionist debate, 1986-1898', *Science & Society*, Vol. 55, No. 2, pp. 232-234.

Harrison, R.T. and Hart, M. (1990), 'The nature and extent of innovative activity in a peripheral regional economy', *Regional Studies*, Vol. 24, No. 5 pp. 383-394.

Harrison, R.T. and Hart, M. (1993), 'New directions in small business research', Avebury, Aldershot.

Harrison, R.T. and Mason, C.M. (1987), 'The regional impact of the small firms loan guarantee scheme in the United Kingdom', *Regional Studies*, Vol. 20, No. 6, pp. 535-550.

Hart, M. (1987), ' The urban labour market impact of new and small manufacturing firms - Some evidence from the Belfast Urban Area', in O'Neill, K., Bhambri, R., Faulkner, T. and Cannon, T. (eds.), *Small business development, some current issues*, Avebury: Aldershot.

Harvey, D. (1988), 'The geographical and geopolitical consequences of the transition from Fordist to flexible accumulation', in Sternliev, G. and Hughes, J. (eds.), *America's new market geography*, Rutgers University Press: New Brunswick.

Harvey, D. (1989), *The condition of postmodernity: An enquiry into the origins of cultural change*, Basil Blackwell: Oxford.

Harvey, P. (1983), 'Marx's theory of the value of labor power: an assessment', *Social Research*, Vol. 50, No. 2.

Hewitt, M.N. (1993), 'Elephants and wallflowers at the global dance: A comparative analysis of the role of training within the economies of Northern Ireland and Massachusetts', Paper presented at the sixteenth national small firms' policy and research conference, UKEMRA: Nottingham.

Hewitt-Dundas, N., McFerran, B. and Roper, S. (1997), 'Competitor knowledge among small firms: A pilot examination of fast growing Northern Ireland companies', *International Small Business Journal*, Vol. 15, No. 2.

Hines, T. (1993), 'Managerial competence and the smaller firm in the UK', Paper presented at the sixteenth national small firms' policy and research conference, UKEMRA: Nottingham.

Hirshhorn, L. (1984), *Beyond mechanization: Work and technology in a post-industrial age*, MIT Press Cambridge: Massachusetts.

Hitchens, D.M.W.N., Birnie, J.E. and Wagner, K. (1989), *Northern Ireland manufacturing productivity compared with West Germany*, NIERC: Belfast.

Hitchens, D.M.W.N. and Birnie, J.E. (1989), 'Economic development in Northern Ireland: Has pathfinder lost its way? A reply', *Regional Studies*, Vol. 23, No. 5, pp. 477-482.

Hitchens, D.M.W.N. and Birnie, J.E. (1992), 'The Scope for Economic Co-operation Between Northern Ireland and the Republic of Ireland: Reversing the Competitive Failure of the Irish Economies', Working Paper, Queen's University of Belfast, Department of Economics.

Hitchens, D.M.W.N. and O'Farrell, P.N. (1988), 'The comparative performance of small manufacturing firms located in the Mid-West of Ireland and Northern Ireland', *Economic and Social Review*, Vol.19, No.3, pp.177-198.

Hitchens, D.M.W.N. and O'Farrell, P.N. (1990), 'Producer services and regional development - Key conceptual issues of taxonomy and quality measurement', *Regional Studies*, Vol. 24, No. 2, pp. 163-171.

Hitchens, D.M.W.N., Wagner, K. and Birnie, J.E. (1990), 'International managerial impressions of the clothing industry: The results of exchange visits by Northern Ireland and German clothing industry managers', Department of Economics Occasional Papers, Queen's University of Belfast: Belfast.

H.M.S.O. (1994), Competitiveness: Helping businesses to win, Cmnd 2563, H.M.S.O.: London.

Hudson, R. (1989), 'Labour market changes and new forms of work in old industrial regions: Maybe flexibility for some, but not flexible accumulation', *Environment and Planning D: Society and Space*, Vol. 7, No. 1, pp. 5-30.

Hymer, S. (1972), 'The multinational corporation and the law of uneven development', in Bhagwati, J. (ed.), *Economics and World Order*, Collier Macmillan: London, pp. 113-140.

Jarvis, V. and Prais, S.J. (1989), 'Two nations of shopkeepers: training for retailing in France and Britain', *National Institute Economic Review*, No. 128, pp. 58-74.

Jessop, B. (1990), *State Theory: putting the capitalist state in its place*, Policy Press: Cambridge, UK.

Jessop, B., Bonnett, K., Bromley, S. and Ling, T. (1988), 'Popular capitalism, flexible accumulation and left strategy', *New Left Review*, Vol. 165, pp. 104-124.

Jurgens, U. (1989), 'The transfer of Japanese management concepts in the international automobile industry', in Wood, S. (ed.), *The transformation of work? Skill, flexibility and the labour process*, Unwin Hyman: London, pp. 204-219.

Kaldor, N. (1972), *The common market*, British and European Economics Series, E2, Holt Information Systems, New York.

Kallenberg, A.L. and Leicht, K.T. (1991), 'Gender and organizational performance: determinants of small business survival and success', *Academy of Management Journal*, No. 34, pp. 136-161.

Katz, H. (1993), 'The decentralization of collective bargaining: A literature review and comparative analysis', *Industrial and Labor Relations Review*, Vol. 47, No. 1, pp. 3-22.

Keeble, D. (1988), 'High technology industry and local environments in the United Kingdom', in Aydalot, P.H. and Keeble, D. (eds.), *High technology industry and innovative environments*, Routledge: London, pp. 65-98.

Keeble, D. (1990), 'Small firms, new firms and uneven regional development in the United Kingdom', *Area - IBG*, Vol. 22, No. 3, pp. 234-245.

Keeble, D. and Egbert, W. (eds.), (1986), *New firms and regional development in Europe*, Croom Helm: London.

Keeble, D. and Kelly, T. (1986), New firms and high technology industry in the United Kingdom: The case of computer electronics, in Keeble, D. and Egbert, W. (eds.), New firms and regional development in Europe, Croom Helm: London, pp. 75-105.

Keeble, D. Owens, P.L. and Thompson, C. (1983), 'The urban-rural manufacturing shift in the European community', *Urban Studies*, Vol. 20, No. 4, pp. 405-418.

Keeble, D. and Tyler, P. (1995), 'Enterprising behaviour and the urban-rural shift', *Urban Studies*, Vol. 32, No. 6, pp. 975-997.

Kelley, M.R. (1989), 'Alternative forms of work organization under programmable automation', in Wood, S. (ed.), *The transformation of work? Skill, flexibility and the labour process*, Unwin Hyman: London, pp. 235-247.

Kirk, W. (1963), 'Problems of geography', *Geography*, No. 48, pp. 357-371.

Kleinknecht, A. and Poot, T. (1992), 'Do regions matter for R&D?', *Regional Studies*, No. 26, pp. 221-32.

Kumazawa, M. and Yamada, J. (1989) 'Jobs and skills under the lifelong nenko employment practice', in Wood, S. (ed.), *The transformation of work? Skill, flexibility and the labour process*, Unwin Hyman: London, pp. 102-127.

Kuratko, D.F. (1988), 'Small business challenging contemporary public policy: A coalition for action', in Judd, R.J., Greenwood, W.T. and Becker, F.W. (eds.), *Small business in a regulated economy*, Quorum: New York, pp.127-139.

Kuznets, S. (1930), *Secular movements in production and prices*, Houghton Mifflin: Boston.

Lampe, D. (1988), *The Massachusetts miracle - High technology and economic revitalization*, Massachusetts Institute of Technology: Cambridge, Massachusetts.

Lash, S. and Bagguley, P. (1988), 'Labour relations in disorganised capitalism: a five nation comparison', *Environment and Planning D: Society and Space*, Vol. 6, No. 3, pp. 321-338.

Lawrence, P. (1980), *Managers and management in West Germany*, Croom Helm: London.

Leborgne, D. and Lipietz, A. (1988), 'New technologies, new modes of regulation - Some spatial implications', *Environment and Planning D: Society and Space*, Vol. 6, No. 3, pp. 263-280.

Lewin, K. (1951), *Field theory in social science; selected theoretical papers*, Harper: New York.

L.E.D.U. (1996), *Measuring the Difference: 1989-1994 LEDU Performance and Impact Study*, Local Enterprise and Development Unit: Belfast.

Lipietz, A. (1986), 'New tendencies in the international division of labour: regimes of accumulation and modes of regulation', in Scott, A.J. and Storper, M. (eds.), *Production, work and territory*, Allen and Unwin: London.

Lipietz, A. (1987), *Mirages and miracles*, Verso: London.

Lipietz, A. (1988), 'Reflections on a tale: The Marxist foundations of the concepts of regulation and accumulation', *Studies in Political Economy*, No. 26, pp. 7-36.

Lowe, J. (1992), 'Locating the line', in Blyton, P. and Turnbull, P (eds.), *Reassessing human resources management*, Sage: London.

Maillat, D. (1991), 'Local dynamism, milieu and innovative enterprises', Longman Cheshire, Melbourne.

Management Charter Initiative, (1987), 'Annual Review', Management Charter Institute, London.

Markusen, A. (1991), 'The military industrial divide', *Environment and Planning D: Society and Space*, Vol. 9, No. 4, pp. 391-416.

Marshall, M. (1987), *Long waves of regional development*, Macmillan: London and Basingstoke.

Mason, G. (1989), 'Some definitional problems in new firm research', *Area*, No. 21, pp. 53-60.

Mason, G., Van Ark, B. and Wagner, K. (1993), 'Productivity, product quality and workforce skills: Food processing in four European countries', *National Institute Economic Review*, Vol. 1, No. 147, pp. 62-83.

Mason, G., Van Ark, B. and Wagner, K. (1994), 'Innovations and the skill mix: chemical and engineering in Britain and Germany', *National Institute Economic Review*, No. 145.

Matthew, R., Wilson, T. and Parkinson, J (1970), *Northern Ireland Development Programme 1970-75*, HMSO: Belfast.

McDermott, P. and Taylor, M. (1982), *Industrial organisation and location*, Cambridge University Press: Cambridge.

Mayo, E. (1949), *The social problems of an industrial civilization*, Routledge: London.
McDonagh, J. (1989), *An overview of vocational education in Massachusetts*, Massachusetts Occupational Outlook: Department of Employment and Training, Boston: Massachusetts, pp. 292-299.
McDonald, R. and Coffield, F. (1991), *Risky business: Youth and the enterprise culture*, Falmer Press: London.
McFerran, B., Hewitt-Dundas, N. and Roper, S. (1996), *Performance benchmarks for developing firms*, Northern Ireland Economic Research Centre: Belfast.
McGregor, D. (1960), *The human side of enterprise*, McGraw-Hill: London.
Meegan, R. (1988), 'A Crisis of Mass Production?' in Allen, J. and Massey, D. (eds.), *The economy in question*, Sage Publications in association with The Open University London: London, pp. 136-183.
Metcalf, D. (1988), Trade unions and economic performance: the British evidence, *Centre for Labour Economics, Discussion paper*, No. 320, London School of Economics, London.
Mintzberg, H. (1975), 'The manager's job: Folklore and fact', *Harvard Business Review*, Vol. 53, pp. 49-61.
Morris, J.L. (1988), 'New technologies, flexible work practices and regional sociospatial differentiation: some observations from the UK', *Environment and Planning D: Society and Space*, Vol. 6, No. 3, pp. 301-319.
Moulaert, F. and Swyngedouw, E. (1989), 'A Regulation approach to the geography of flexible production systems', *Environment and Planning D: Society and Space*, Vol. 7, No. 3, pp. 327-345.
Mullen, J., Armstrong, J. and Kavanagh, J. (1986), 'From mill town to mill town: the transition of a New England town from a textile to a high technology economy', *Journal of the American Planning Association*, Vol. 52, pp. 47-59.
Murie, A.S., Birrell, W.D., Hillyard, P.A.R. and Roche, D.J.D. (1974), 'Regional policy and the attraction of manufacturing industry in Northern Ireland', Working Paper, Centre for Environmental Studies: London.
Murray, P. (1988), 'Karl Marx as a historical materialist historian of political economy', *History of political economy*, Vol. 20, No. 1, pp. 95-106.
Murray, R. (1988), 'Life after Henry Ford', *Marxism Today*, October, pp. 8-13.
Murray, R. (1989), 'Interview in interlinks', *Journal of the Socialist Society*, February-March, pp. 11-14.
Murshed, S.M., Noonan, D., Thanki, R., Gudgin, G. and Roper, S, (1993), 'Growth and development in the two economies of Ireland: An overview', Working Paper No. 5, Northern Ireland Economic Research Centre: Belfast.

N.I.E.C. (1983), 'Economic strategy: Historical growth performance', Northern Ireland Economic Council Report, No. 38, Belfast.
N.I.E.C. (1983), 'The duration of industrial development maintained employment', Northern Ireland Economic Council Report, No. 39, Belfast.
N.I.E.C. (1990), 'The Industrial Development Board for Northern Ireland: Selective financial assistance and economic development policy', Northern Ireland Economic Council Report, No. 79, Belfast.
N.I.E.C. (1991), 'Economic strategy in Northern Ireland', Northern Ireland Economic Council Report, No. 88, Belfast.
National Economic Development Office (NEDO) (1986), *Changing working patterns: How companies achieve flexibility to meet new needs*, National Economic Development Office: London.
Nelson, R.R. and Winter, S.G. (1982), 'The Schumpeterian tradeoff revisited', *American Economic Review*, Vol. 62, No. 1, pp. 114-132.
Nielsen, K. (1991), 'Towards a flexible future - theories and politics', in Jessop, B., Kastendiek, H., Nielsen, K., and Pedersen, O.K., *The politics of flexibility - Restructuring state and industry in Britain, Germany and Scandanavia*, Edward Elgar: New York, pp. 34-54.
NIERC (1988), *Job Generation and Manufacturing Industry, 1973-86*, Northern Ireland Economic Research Centre: Belfast.
Nolan, P. and O'Donnell, K. (1991), *Restructuring and the politics of industrial renewal - the limits of flexible specialization*, Basil Blackwell: Oxford.
North, D. and Smallbone, D. (1993), 'Employment generation and small business growth in different geographical environments', Paper presented at the sixteenth national small firms' policy and research conference, UKEMRA: Nottingham.
Northern Ireland Annual Abstract of Statistics, (1983), Policy Planning and Research Unit, Belfast.
Northern Ireland Growth Challenge (1995), *Interim Summary of Progress*, Northern Ireland Growth Challenge: Belfast.
O'Neill, K. and MacNabb, A. (23/2/94), 'Small minded', *Business News Letter*, p.14.
Oakey, R. P. (1984), *High Technology Small Firms*, Pinter Francis, London.
Oberhauser, A.M. (1990), 'State policy, employment and the spatial-organization of production - The establishment of an automobile plant in the North of France', *Tijdschrift Voor Economische en Sociale Geografie*, Vol. 81, No. 1, pp. 58-68.
Office of Technology Assessment (OTA) (1984), *Worker training: Competition in the new international economy*, OTA: Boston, Massachusetts.
Office of Technology Assessment (OTA), (1990), *Making Things Better, Competing in Manufacturing,* OTA: Boston, Massachusetts.

Palloix, C. (1976), 'The labour process: From Fordism to Neo-Fordism', in CSE Pamphlet, *The labour process and class strategies*, Stage 1 Conference for Socialist Economists, pp. 46-67.

Papadimitriou, Z. (1986), 'Changing skill requirements and trade union bargaining', in Jacobi, O., Jessop, B., Kastendiek, H. and Regini, M. (eds.), *Technical change, rationalisation and industrial relations*, Croom Helm: London, pp. 35-49.

Peck, J.A. and Tickell, A. (1991), 'Regulation theory and the geographies of flexible accumulation: Traditions in capitalism, transitions in theory', SPA Working Paper 12, School of Geography, University of Manchester.

Peck, J.A. and Tickell, A. (1992), 'Time, space and flexibility: The social regulation of uneven development', Paper presented to the conference of the International geographical union commission on industrial change, Time, space, competition and contemporary industrial change, Orlando.

Pinch, S., Mason, C. and Witt, S. (1991), 'Flexible employment strategies in British industry - Evidence from the UK sunbelt', *Regional Studies*, Vol. 25, No. 3, pp. 207-218.

Piore, M. and Sabel, C.F. (1984), *The second industrial divide*, Basic Books: New York.

Pollert, A. (1988), 'From core-periphery to flexible specialization', Paper presented at the conference on the labour process, Aston, UMIST.

Pollert, A. (1991), *Farewell to flexibility? Questions of restructuring*, Basil Blackwell: Oxford.

Porter, M.E. (1980), *Competitive Strategy*, The Free Press-Division of Macmillan: London.

Porter, M. (1990), The Competitive Advantage of Nations, The Free Press, New York.

Prais, S.J. (1981), 'Vocational qualifications of the labour force in Britain and Germany', *National Institute Economic Review*, No. 98, pp. 47-59.

Prais, S.J. (1986), 'Some international comparisons of the age of the machine stock', *Journal of Industrial Economics*, No. XXXIV, pp. 261-277.

Prais, S.J. (1989), 'Qualified manpower in engineering - Britain and other industrially advanced countries', *National Institute Economic Review*, No. 127, pp. 76-84.

Prais, S.J., Jarvis,V. and Wagner,K. (1989), 'Productivity and vocational skills in Britain and Germany: Hotels', *National Institute Economic Review*, No. 130, pp. 52-74.

Prais, S.J. and Wagner, K. (1983), 'Some practical aspects of human capital investment', *National Institute Economic Review*, No.105, pp. 46-65.

Prais, S.J. and Wagner, K. (1985), 'Schooling standards in England and Germany: Some summary comparisons bearing on economic performance', *National Institute Economic Review*, May.

Rajan, A. and Pearson, R. (eds.) (1986), *UK occupation and employment trends to 1990*, Butterworth: London.

Reynolds, P.D. and Miller, B. (1988), *1988 Minnesota new firms study: An exploration of new firms and their economic contributions*, Centre for Urban and Regional Affairs, University of Minnesota: Minneapolis, Minnesota.

Roche, F. and Tansey, P. (1992), 'Industrial training in Ireland', Report to the Industrial Policy Review Group, Dublin.

Roper, S., Ashcroft, B., Love, J.H., Dunlop, S., Hofmann, H. and Vogler-Ludwig, K. (1996), *Product innovation and development in the UK, German and Irish firms*, Northern Ireland Economic Research Centre and Fraser of Allander: Belfast.

Roper, S., Hewitt-Dundas, N. and McFerran, B. (1997), 'Disparities in quality perceptions between small firms and their customers', *International Small Business Journal, (Forthcoming)*.

Roper, S. (1994), 'Training, skills and competitiveness: Northern Ireland compared with Germany', Paper presented at the conference on Industrial policy, competitiveness and growth, Northern Ireland Economic Research Centre: Belfast.

Roper, S. and Hofmann, H. (1993), *Training, skills and company competitiveness: A comparison of matched plants in Northern Ireland and Germany*, Northern Ireland Economic Research Centre: Belfast.

Rosegrant, S. and Lampe, D. (1992), *Route 128 - Lessons from Boston's high tech community*, Basic Books: New York.

Ross, R.J.S. and Trachte, K.C. (1990), *Global capitalism: The new leviathan*, State University of New York Press: Albany.

Rothwell, R. (1982), 'The role of technology in industrial change: implications for regional policy', *Regional Studies*, Vol.16, No. 5, pp. 361-370.

Sabel, C. (1982), *Work and politics: The division of labour in industry*, Cambridge University Press: Cambridge.

Sabel, C. (1989), 'The reemergence of regional economies', in Hirst, P. and Zeiltin, J. (eds.), *Reversing industrial decline?*, The Berg Press: Leamington Spa, pp.17-70.

Sayer, A. (1985), 'Industry and space: A sympathetic critique of radical research', *Society and Space*, Vol. 3, No. 1, pp. 3-29.

Sayer, A. and Walker, R. (1992), *The new social economy - Reworking the division of labour*, Blackwell: Oxford.

Schoenberger, E. (1982), Japanese Manufacturing Techniques, The Free Press: New York.

Schoenberger, E. (1988), 'From Fordist to flexible accumulation - Technology, competitive strategies, and international location', *Environment and Planning D: Society and Space*, Vol. 6, No. 3, pp. 245-262.

Schumpeter, J. (1934), *The theory of economic development*, Harvard University Press, Cambridge: Massachusetts.

Schumpeter, J. (1939), *Business cycles*, McGraw-Hill: New York.

Schumpeter, J. (1961), *The theory of economic development*, Oxford University Press: Oxford.

Scott, A. (1983), 'Industrial organization and the logic of intra-metropolitan location: theoretical considerations', *Economic Geography*, Vol. 59, No. 3, pp. 233-250.

Scott, A. and Storper, M. (1988), 'The geographical foundations and social regulation of flexible production complexes', in Wolch, J. and Dear, M. (eds.), *Territory and social reproduction*, Allen & Unwin: Boston.

Scott, A.J. (1988), 'Flexible production systems and regional development: the rise of new industrial spaces in North America and Western Europe', *International Journal of Urban and Regional Research*, No. 12, pp. 171-186.

Sengenberger, W., Loveman, G.W. and Piore, M.J. (1990), *The re-emergence of small enterprises: Industrial restructuring in industrialised countries*, International Labour Organisation: Geneva.

Senker, J. (1988), *A taste for innovation, British supermarkets' influence on food manufacturers*, Horton: Bradford.

Shaiken, H., Herzenberg, S. and Kuhn, S. (1986), 'The work process under more flexible production', *Industrial Relations*, Vol. 25, No. 2, pp. 167-183.

Shimizu, R. (1990), 'Strategy of competitive advantage and network of strengths in large Japanese companies', *KEIO Business Review*, No. 27, Society of Business and Commerce: KEIO University.

Shutt, J. and Whittington, R. (1987), 'Fragmentation strategies and the rise of small units - cases from the North-West', *Regional Studies*, Vol. 21, No.1, pp. 13-23.

Small Business Administration (1991), *The state of small business: A report of the President*, US Government Printing Office: Washington, D.C.

Smallbone, D. (1988), 'Enterprise agencies and the survival of new small business start ups', *Local Economy*, pp.143-47.

Smallbone, D., North, D. and Leigh, R. (1993), ' The growth and survival of mature manufacturing SMEs in the 1980s: An urban and rural Comparison', in Curran, J. and Storey, D.J. (eds.), *Small firms in urban and rural locations*, Routledge: London.

Smith, A.D., Hitchens, D.M.W.N. and Davies, S.W. (1982), 'International industrial productivity - A comparison of Britain, America and Germany', *NIESR Occasional Papers*, No. XXXIV, Cambridge University Press: Cambridge.

Smith, A.D. Hitchens, D.M.W.N. and Davies, S.W. (1982), *International Industrial Productivity*, Cambridge University Press: Cambridge, England.

Solem, O. and Steiner, M.P. (1989), 'Factors for success in small manufacturing firms - And with special emphasis on growing firms', in Virtanen (ed.), *Proceedings of the conference on SME's and the Challenges of 1992*, Helsinki School of Economics: Helsinki.

Starr, E. (1988), 'The growth of small manufacturers: 1976-1984', *Business Economics*, Vol. 23, pp. 41-45.

Steedman, H. and Wagner, K. (1987), 'A second look at productivity, machinery and skills in Britain and Germany', *National Institute Economic Review*, No. 122, pp. 84-95.

Steedman, H. and Wagner, K. (1989), 'Productivity, machinery and skills: clothing manufacturing in Britain and Germany', *National Institute Economic Review*, No. 128, pp. 40-57.

Storey, D.J. and Johnson, S. (1987), *Job generation and labour market change*, Macmillan: London.

Storey, D.J., Keasey, K., Watson, R. and Wynarczyk, P. (1987), *The performance of small firms: profits, jobs and failures*, Croom Helm: London.

Storey, D. (1994), *Understanding the small business sector*, Routledge: London.

Storey, D.J and Strange, A. (1991), 'New players in the enterprise culture', Paper presented at the sixteenth national small firms' policy and research conference, UKEMRA, Blackpool.

Storey, D.J. (1982), *Entrepreneurship and the new firm*, Croom Helm: London.

Storey, D.J. (1985), 'The problems facing new firms', *Journal of Management Studies*, Vol. 22, No. 3, pp. 327-345.

Storper, M. and Walker, R. (1989), *The capitalist imperative - territory, technology and industrial growth*, Basil Blackwell: New York.

Streeck, W. (1987), 'The uncertainties of management in the management of uncertainty: employers, labor relations and industrial adjustment', *Work Employment and Society*, Vol.1, No. 3, pp. 281-308.

Sugimori, Y., Kusunoki, F., Cho, F. and Uchikawa, S. (1977), 'Toyota production system and kanban system: Materialization of just-in-time and respect-for-human system', *International Journal of Production Research*, Vol. 15, No. 6, pp. 553-564.

Swyngedouw, E.A. (1991), 'The production of new spaces', unpublished doctoral thesis, The John Hopkins University.

Taylor, W. (1971), 'A secondary curriculum for the average child', in Hooper, R. (ed.), *The curriculum: Context, design and development*, Unwin Educational Books: London, pp. 49-61.

Telesis (Report) (1982), *Review of Industrial Policy*, National Economic and Social Council Report, No. 64, Dublin.

Thwaites, J. (1983), 'Trade unions in Canada - 1812-1902', *Industrial Relations*, Vol. 38, No. 1, pp. 179-180.

Tomaney, J. (1990), 'The reality of workplace flexibility', *Capital and Class*, No. 40, Spring 1990, pp. 29-61.

Townsend, A.R. (1983), 'The Impact of the recession: on industry, employment and the regions, 1976-1981', Croom Helm, London-Canberra.

Vaessen, P. and Keeble, D. (1995), 'Growth oriented SMEs in unfavourable regional environments', ESRC Centre for Business Research, Working Paper No. 6, University of Cambridge: Cambridge.

Walker, R. (1989), 'Machinery, labour and location', in Wood, S (ed.) *The transformation of work? Skill, flexibility and the labour process*, Unwin Hyman: London, pp. 59-91.

Watts, H.D. (1987), *Industrial Geography*, Longman: New York.

Westhead, P. and Birley, S. (1993), 'Employment growth in new independent owner-managed firms in Great Britain', Paper presented at the Seventh European Entrepreneurship Research Conference - Rent VII, Budapest, Hungary.

Wilson Committee (1979), *The financing of small firms, interim report of the committee to review the functioning of the financial institutions*, Cmnd 7503, HMSO: London.

Woo, C.Y., Cooper, A.C., Dunkelberg, W.C., Dallenbach, U. and Dennis, W.J. (1989), 'Determinants of growth for small and large entrepreneurial startups', in Brockhaus, R. H., Churchill, M.C., Katz, J.A., Kirchhoff, B.A., Vesper, K.H. and Wetzel, W.E. (eds.), *Frontiers Of Entrepreneurship*, Babson College, Boston: Massachusetts, pp. 134-147.

Wood, S. (ed.) (1989), *The transformation of work? Skill, flexibility and the labour process*, Unwin Hyman: London.

Woods, A.., Blackburn, R. and Curran, J. (1993), *A longitudinal study of small enterprises in the service sector*, Small Business Research Centre, Kingston University and Department of Management Studies, Brunel University.

Wynarczyk, P., Watson, R., Storey, D.J., Short, H. and Keasey, K. (1993), *Managerial labour markets in small and medium sized enterprises*, Routledge: London.